Joining and Welding Symposium 2017

IOP Conference Series: Materials Science and Engineering
Volume 238

Pekan, Malaysia
11 July 2017

ISBN: 978-1-5108-5197-9
ISSN: 1757-8981

Printed from e-media with permission by:

Curran Associates, Inc.
57 Morehouse Lane
Red Hook, NY 12571

Some format issues inherent in the e-media version may also appear in this print version.

Printed by Curran Associates, Inc. (2017)

For permission requests, please contact the Institute of Physics at the address below.

Institute of Physics
Dirac House, Temple Back
Bristol BS1 6BE UK

Phone: 44 1 17 929 7481
Fax: 44 1 17 920 0979

techtracking@iop.org

Additional copies of this publication are available from:

Curran Associates, Inc.
57 Morehouse Lane
Red Hook, NY 12571 USA
Phone: 845-758-0400
Fax: 845-758-2633
Email: curran@proceedings.com
Web: www.proceedings.com

Table of contents

Volume 238

Joining and Welding Symposium
11 July 2017, Universiti Malaysia Pahang, Malaysia

Accepted papers received: 4 September 2017
Published online: 23 October 2017

Preface

Papers

IOP Conf. Series: Materials Science and Engineering **238** (2017) 011001 doi:10.1088/1757-899X/238/1/011001

SECTION 3: PRELIMINARY MATERIAL

PREFACE

Joining and Welding Symposium 2017 is a one-day symposium broadly targeted to gather experts, professionals and researchers from UMP in related areas of soldering, brazing and welding to share recent advances and development accomplished.

The aim of this symposium is to provide a unique platform for interactive exchange of latest state-of-the-art knowledge among the professionals in soldering, brazing and welding domain. In addition, the symposium will be a networking event to spur future collaborations between researchers of UMP.

SPONSOR/FUNDING ACKNOWLEDGEMENTS

Universiti Malaysia Pahang

LIST OF COMMITTEES

No	Name	Affiliation
1	Professor Dr. Rizalman Mamat	Universiti Malaysia Pahang
2	Profesor Madya Dr. Mahadzir Ishak	Universiti Malaysia Pahang
3	Dr Siti Rabiatull Aisha binti Idris	Universiti Malaysia Pahang
4	Dr Aiman bin Mohd Halil	Universiti Malaysia Pahang
5	Dr Nur Zulikha Mohd Khalil	Universiti Malaysia Pahang
6	Nabila Tamar Jaya	Universiti Malaysia Pahang
7	Muhammad Zakwan Che Nik	Universiti Malaysia Pahang
8	Khairul Ihsan Yaakob	Universiti Malaysia Pahang
9	Muhammad Naqiuddin Mat Salleh	Universiti Malaysia Pahang
10	Muhammad Izzat Izzuddin Mahamud	Universiti Malaysia Pahang
11	Mohamed Syafiq Khuzairi Yaacob	Universiti Malaysia Pahang

IOP Conf. Series: Materials Science and Engineering **238** (2017) 011001 doi:10.1088/1757-899X/238/1/011001

IOP Conf. Series: Materials Science and Engineering **238** (2017) 011001 doi:10.1088/1757-899X/238/1/011001

IOP Conf. Series: Materials Science and Engineering **238** (2017) 011001 doi:10.1088/1757-899X/238/1/011001

IOP Conf. Series: Materials Science and Engineering **238** (2017) 011001 doi:10.1088/1757-899X/238/1/011001

Main topics:

IOP Conf. Series: Materials Science and Engineering **238** (2017) 011001 doi:10.1088/1757-899X/238/1/011001

IOP Conf. Series: Materials Science and Engineering **238** (2017) 011001 doi:10.1088/1757-899X/238/1/011001

IOP Conf. Series: Materials Science and Engineering **238** (2017) 011001 doi:10.1088/1757-899X/238/1/011001

IOP Conf. Series: Materials Science and Engineering **238** (2017) 011001 doi:10.1088/1757-899X/238/1/011001

IOP Conf. Series: Materials Science and Engineering **238** (2017) 011002 doi:10.1088/1757-899X/238/1/011002

Peer review statement

All papers published in this volume of *IOP Conference Series: Materials Science and Engineering* have been peer reviewed through processes administered by the proceedings Editors. Reviews were conducted by expert referees to the professional and scientific standards expected of a proceedings journal published by IOP Publishing.

IOP Conf. Series: Materials Science and Engineering **238** (2017) 012001 doi:10.1088/1757-899X/238/1/012001

Tin Whiskers Formation and Growth on Immersion Sn Surface Finish under External Stresses by Bending

N L M Amin, S Z Yusof, M N A Kahar, T A A Bakar, A O and N A Fadil

Department of Materials, Manufacturing, and Industrials Engineering, Faculty of Mechanical Engineering, Universiti Teknologi Malaysia, 81310 Skudai, Johor Bahru, Malaysia

E-mail: norakmal@fkm.utm.my

Abstract. Deposited tin (Sn) layers used as lead-free solderable finish on the semiconductor devices are known to form whiskers. These whiskers are a single crystal of tin that spontaneously grows from the surface of tin within weeks to years. Thus, they can cause shorts and the failure of a whole electronic circuit. In this research, the effect of external stresses applied by bending on the tin whiskers formation and growth was investigated on immersion Sn surface finish. The plating time was 10 minutes to produce 1.4 μm coating thickness on the copper substrate and exposed for 1, 4, and 8 weeks under the environment of 30°C/60% RH for bent and non-bent samples. It was found that the non-bent Sn surface had uniform ts distributed on the entire surface. Unlike the non-bent Sn surface, the surface of bent samples had non-uniform tin whiskers distribution. The tin whiskers formed more and grew longer at the lower stress region of the bent surface as compared to the higher stress region, based on the micrograph observed using the field emission scanning electron microscopy (FESEM).

1. Introduction

The initiative of removing lead (Pb) from the materials set within the soldering technology in microelectronics industry has created many challenges in research and development. One of the issues is to replace Pb with other suitable materials. Pure tin (Sn) and alloys with high Sn content have been the leading choices for many manufacturers. Pb was deemed a toxic material in electronic components and was subsequently banned, making pure Sn the right substitute due to its good conductivity [1], good wettability [2], and low cost [3]. These Sn finishes, however, have the tendency to spontaneously grow whiskers on the surface, which is considered by some to be a potential reliability concern as the whiskers can continuously grow and may cause shorting between components. Tin whiskers on Sn surface finish studies have been reported by many researchers [2,4-10] as it has become a major concern in optimizing the use of Pb-free Sn alloys in the electronic industry. Most of the previous studies focused on whisker growth on electroplated tin and almost none used immersion tin plating.

The main purpose of this research is to investigate and understand the effects of external stress under controlled environmental storage conditions such as temperature and humidity on tin whiskers formation and growth, specifically on immersion Sn surface finish since deposition of Sn on Cu using immersion process is a residual stress-free coating. Externally applied stress will be induced by bending to simulate the effect of stress on the formation and growth of whiskers. In this study, as tin whiskers growth is mostly a stress related phenomenon, the findings of Sn behaviour were used to interpret the process of stress build up and relief in both non-bent (as-plated) and bent Sn. The study is taken to understand the mechanism of tin whiskers growth when external stress is applied through the bending process of the coated materials.

IOP Conf. Series: Materials Science and Engineering **238** (2017) 012001 doi:10.1088/1757-899X/238/1/012001

2. Experimental Procedure

Whisker growth has been observed on a number of different substrate materials commonly used in the electronics industry. The substrate material used in the experiment was a commercial oxygen-free pure copper with 1 mm thickness, 35 mm width and 50 mm length. Plating process by immersion was conducted by following the method produced by Fadil et al. [11]. Plating bath equipped with stainless steel heating coil was set up for the plating process in order to maintain the temperature of plating. The plating bath was prepared by dissolving 20 g/L of stannous chloride as a tin precursor, 16 g/L of sodium hypophosphite as a reducing agent, 37 ml/L of hydrochloric acid (37% ml), 50 ml/L of Sulfuric acid (50% ml), 200 g/L of thiourea, and 5 ml/L of phenolsulfonic acid in 1L distilled water. The immersion tin process was conducted at temperature of 75 °C and the deposition time was 10 minutes in order to obtain 1.4 μm of tin thickness.

Reflow process was performed in a reflow oven with the peak reflow temperature set at 260°C freshly after the coating process. After the reflow process, the samples were slowly bent to 45° until a crack occurred. The bending process was conducted by following the JEDEC standard, IPC/JEDEC-9702 [12]. Two sets of samples; bent and non-bent samples (Figure 1) were then stored in the humidity chamber for time intervals of 1, 4, and 8 weeks under the environment of 30°C/60% RH. The surface morphologies of the bent samples were examined using field-emission scanning electron microscopy (FESEM) at three different locations; the edge (farthest location from the crack), crack, and center (region between the edge and the crack).

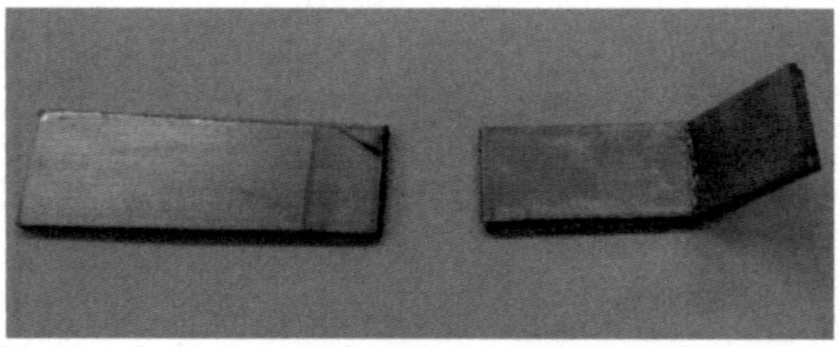

Figure 1. Photo of as prepared sample (left) and bent sample (right).

3. Results and Discussion

Figure 2 shows the whiskers formation for non-bent samples exposed under the environment of 30°C/60% RH for 1 week (Figure 2.a), 4 weeks (Figure 2.b), and 8 weeks (Figure 2.c). From Figure 2, it can be clearly seen that the whiskers have grown longer over exposure time. Non-bent samples were observed at the entire surface and the observation found was that the formation of whiskers were considerably uniform at every area on the samples. This is due to the uniformly distributed stress experienced on the sample rather than the stress caused by bending, which would result in higher stress at the crack region and lower stress at the outer crack region.

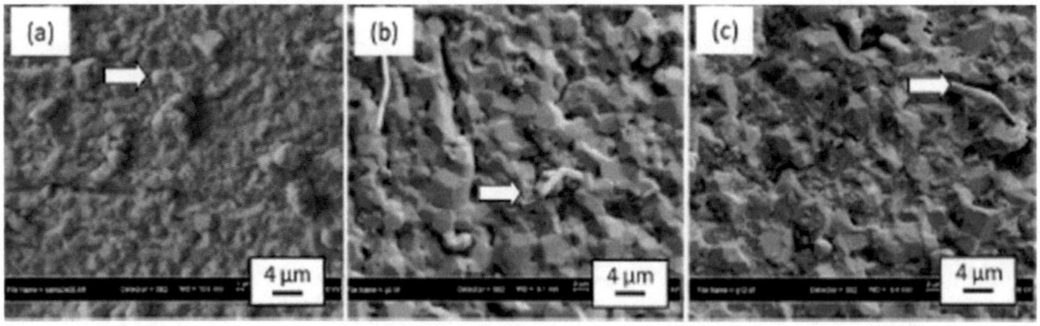

Figure 2. FESEM micrographs of tin whiskers formed on the non-bent surface exposed for (a) 1 week, (b) 4 weeks, and (c) 8weeks under 30°C/ 60% RH.

Figure 3, 4 and 5 show the whiskers formation on the bent samples at selected areas exposed for 1 week, 4 weeks, and 8 weeks respectively. Three different regions were examined under FESEM which are the edge (the region farthest from the bent point), the centre (the region between the edge and the bent point), and the crack region (the bent point itself). FESEM observation found that more whiskers formed at the edge area than the crack area after 1 week (Figure 3), 4 weeks (Figure 4) and 8 weeks (Figure 5) exposed under the environment of 30°C/60% RH. This was due to the higher stress concentration at the crack area created by bending.

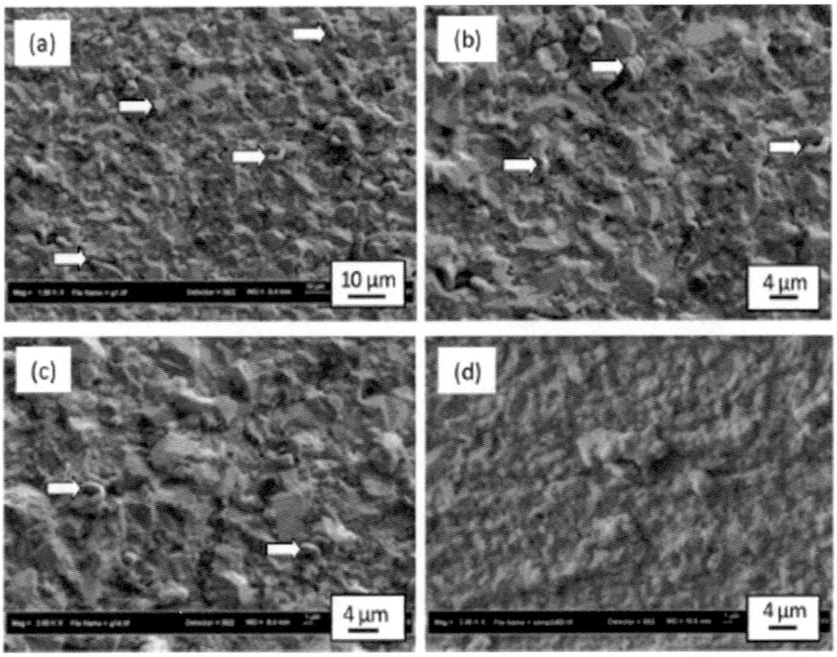

Figure 3. FESEM micrographs of whiskers formed at the (a) edge (low magnification), (b) edge (high magnification), (c) centre, and (d) crack area for bent surface after 1 week exposed under 30°C/ 60% RH.

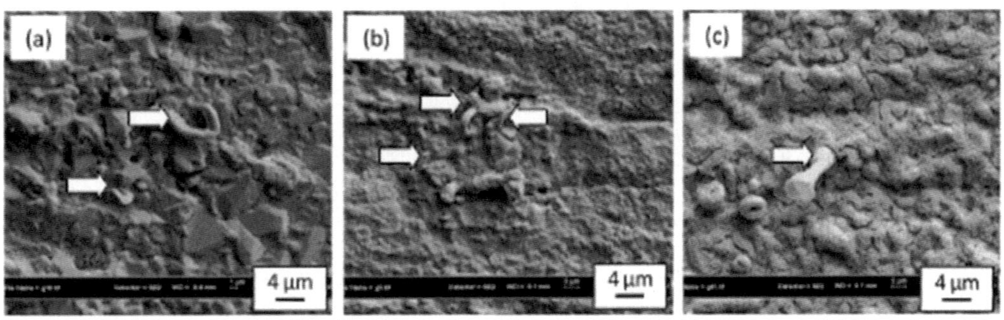

Figure 4. FESEM micrographs of whiskers formed at the (a) edge, (b) centre, and (c) crack area for bent surface after 4 week exposed under 30°C/ 60% RH.

Figure 5. FESEM micrographs of whiskers formed at the (a) edge, (b) centre, and (c) crack area for bent surface after 8 week exposed under 30°C/ 60% RH.

The whiskers formation and growth behavior can be explained clearly by the schematic diagram in Figure 6. The numbers of whiskers formed at the bent area were less than the outside region. This is because the outer region has lower stress compared to the crack area. It should be understood that the bent samples have higher stress concentrated at the crack and stress decreases towards the farther region from the crack area. This resulted in the migration of Sn atoms from higher to lower stress region to release the stress [13].

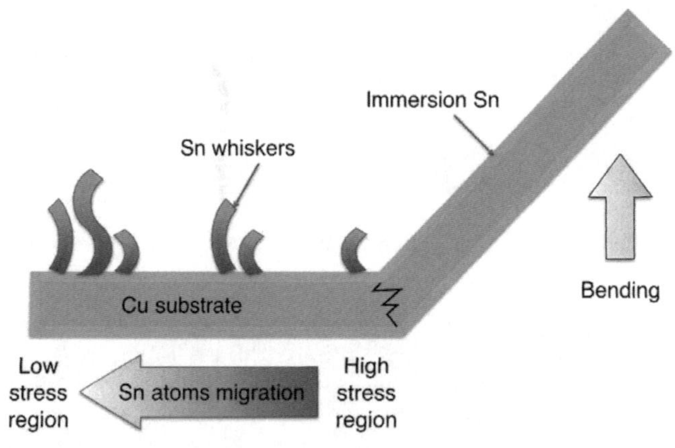

Figure 6. Schematic diagram of stress effect on tin whiskers formation for immersion Sn surface finish.

4. Conclusion

The effect of externally applied stress by bending caused the least tin whiskers formation at the crack region, more at the centre region, and the most tin whiskers formation at the edge region of the Sn surface finish. This is because the crack area has a high stress concentration while the centre and edge has lower stress concentration. The whiskers formation increases with lower stress as Sn atoms migrate from higher to lower stress concentration region to release the stresses. Unlike the bent Sn surface, the non-bent Sn surface has uniform tin whiskers distributed on the entire surface.

Acknowledgement

This work was supported by the team from Solder Metallurgy Laboratory and Materials Science Laboratory, Faculty of Mechanical Engineering, Universiti Teknologi Malaysia, and the Ministry of Education Malaysia and through the FRGS Grant (Grant. No. R.J130000.7824.4F661) and UTM Research Grant (GUP Tier 1: Q.J130000.2524.13H72).

References

[1] Dimitrovska A and Kovacevic R 2009 Mitigation of tin whiskers growth by composite Ni/Sn plating *J. Electron. Mater.* **38** 2516-2524

[2] Vicenzo A 2011 *Tin Whiskers. In: C. Z.G. GROSSMANN, (ed.) The ELFNET Book on Failure Mechanisms, Testing Methods, and Quality Issues of Lead-Free Solder Interconnects* (Springer-Verlag London)

[3] Illes B, Horvath B, and Harsanyi G 2010 Effect of strongly oxidizing environment on whisker growth form tin coating *Surf. Coat. Technol*, **205** 2262-2266

[4] Heng Z et al. 2017 The application and research status of tin whisker formation in electric usage *AIP Conference Proceedings* **1839** 020003

[5] Ashworth M A and Dunn B 2016 An investigation of tin whisker growth over a 32-year period *Circuit World* **42**(4) 183-196

[6] Singh S S et al. 2014 Tensile behavior of single-crystal tin whiskers *J. Electron. Mater.* **43** (4) 978–982

[7] Balazs I and Barbara H 2014 Tin whisker growth from micro-alloyed SAC solders in corrosive climate *Journal of Alloys and Compounds* **616** 116-121

[8] Puttlitz K J and Galyon G T 2007 Impact of the ROHS directive on high-performance electronic systems *J Mater. Sci: Mater. Electron.* **18** 347-365

[9] Baated A, Kim K S, and Suganuma K 2011 Effect of intermetallic growth rate on spontaneous whisker growth from a tin coating on copper *J Mater. Sci: Mater. Electron.* **22** 1685-1693

[10] Miller S M, Sahaym U, and Norton M G 2010 Effect of substrate composition on tin whiskers growth in pure Sn films *Metall Mater Trans A* **41 A** 3386-3395

[11] Fadil N A, Yusof S Z, Ourdjini A, Abu Bakar T A 2015 Effect of the formation of CuO flowers and SnO_2 on the growth of tin whiskers on immersion tin surface finish *Applied Mechanics and Materials* **695** 301-305

[12] JEDEC 2004 *Monotonic bend characterization of board-level interconnects* IPC/JEDEC-9702

[13] Shibutani T, Yu Q 2008 Pressure Induced Tin Whisker Formation on SnCu Finish by Nanoindentation Creep *The proceeding of 10th Electronic Packaging Technology Conference* 1442-1447

IOP Conf. Series: Materials Science and Engineering **238** (2017) 012002 doi:10.1088/1757-899X/238/1/012002

Study of Dissimilar Welding AA6061 Aluminium Alloy and AZ31B Magnesium Alloy with ER5356 Filler Using Friction Stir Welding

M I I Mahamud, M Ishak and A M Halil

Faculty of Mechanical Engineering, Universiti Malaysia Pahang, 26600, Pekan, Pahang, Malaysia

E-mail: izzat_2377@yahoo.com

Abstract. This paper is to study of dissimilar welding AA6061 aluminium alloy and AZ31B magnesium alloy with ER5356 filler using friction stir welding. 2 mm thick plates of aluminium and magnesium were used. Friction stir welding operations were performed at different rotation and travel speeds and used the fixed tilt angle which is 3°. The rotation speeds varied from 800 to 1100 rpm, and the travel speed varied from 80 to 100 mm/min. In the range rotation speed of 800 to 1000 rpm and welding speed of 80 to 100 mm/min there are no defect at the weld. Tensile test show the higher tensile strength is 198 MPa and the welding efficiency is about 76%.

1. Introduction

Magnesium alloys do not have enough strength to apply as a structural material although it is lightweight, combination of aluminium alloy and magnesium alloy can be a good structure. However, conventional fusion welding between aluminium alloy and magnesium alloy is unable because large coefficient of expansion of $Mg_{12}Al_{17}$ intermetallic compound formation in the fusion zone [1]. The use of conventional arc welding is unable because it will produce large HAZ because of melting of the base metal. It will produce defect due to high heat input and slow cooling rate and difficult to joint dissimilar material because of different chemical composition such as melting point. To overcome the disadvantages, Friction Stir Welding is an alternative method. Friction Stir Welding is a method process occurs below than the melting point of the alloys that is solid state joining.

Friction stir welding is one of the solid state welding. This method uses a cylindrical shouldered tool with a profiled pin. The pin is rotating while in the line of the two joining work piece that clamped at the backing plate [2].it use thermal heating by mechanical stirring by the rotational tool [3]. This welding is not achieving melting temperature of base metal. It is advances in aerospace, automotive and ship building industrial. Besides, multiple aspects of robotic friction stir welding are covered, including sensing, control and joint tracking [4]. The advantages in production by FSW are energy efficient, environment friendly and versatile. Then, FSW can be used to joining high strength aerospace aluminium alloys and other metallic alloy that are hard to weld by conventional fusion welding [3]. FSW produce the weld which joint metal without melting the base metal, improve in strength, ductility, fatigue and fracture toughness. In addition, weld by FSW achieved 80% yield strength of base metal and good appearance weld and low distortion [5]. This method uses a cylindrical shouldered tool with a profiled pin. The pin is rotating while in the line of the two joining work piece that clamped at the backing plate.it use thermal heating by mechanical stirring by the rotational tool. This welding is not achieving melting temperature of base metal [6].

Then, intermetallic compound layer in the stir zone is lower than that in the fusion zone of fusion welding [7]. Friction Stir welding can control the brittle intermetallic compound. There are some of researchers that conduct the experiment of dissimilar FSW between AZ31B and AA6061. Malarvizhi and Balasubramanian 2012 has conducted FSW between AZ31B and AA6061 joints were evaluated using tools with different shoulder diameters, the joint fabricated with a tool shoulder diameter of 21 mm (3.5 times the plate thickness) yielded maximum tensile strength of 192 MPa and the joint efficiency is 89% [8]. Other, Masoudian, Tahaei et al. 2014 has also conducted Tensile strength of the welded specimen was about 76% of that of AZ31 Mg alloy and AA6061 Al alloy, and the welded specimen failed through brittle-mode fracture [9]. Then, Morishige, Kawaguchi et al. 2008, conducted the dissimilar FSW joint between A5052-H aluminium alloy and AZ31B magnesium alloy was able to join and the joint efficiency was achieved to 61% [1].

The improvement on efficiency by enhance intermetallic compound by using filler. Then, to study what is effect of joining using filler because no study uses filler in FSW. Lack of studies observed in joining dissimilar AZ31B magnesium alloy and aluminium AA6061. It is difficult to obtain the optimum parameter such as tool rotational speed and transverse speed make the reason to run the experiment.

2. Experimental Procedure

The materials that are used are AA6061 and AZ31B with thickness 2 mm. Then, it cut into 100mm x 60mm dimension. Table 1 and table 2 show the chemical compositions of both materials.

Table 1.

Material	Mg	Si	Fe	Cr	Cu	Mn	Al
AA6061	0.89	0.53	0.20	0.20	0.30	0.02	Balance

Chemical composition (wt. %) of AA6061 aluminium alloy.

Material	Al	Zn	Mn	Si	Cu	Ca	Mg
AZ31B	3.0	1.00	0.34	0.02	0.02	0.003	Balance

Table 2. Chemical composition (wt. %) of AZ31B magnesium alloy.

In this friction stir welding process. The plates were cleaned with a sand paper to remove surface oxides. Then, the plates cleaned with acetone to cleaning the surface of the plate. Unthreaded tapered tool made from H13 hardened tool steel with a shoulder of 10 mm in diameter an, a pin of 2 mm in diameter and 1.7 mm in length, and a tilt angle of 3° was used during the welding. Aluminium rod is used in this experimental which is ER5356 aluminium filler. Aluminium rod filler was knocked into flat 2mm × 0.3mm size by using hammer to require the dimension. In this experiment, the parameters used are 800rpm-100rpm rotation speed and 80 mm/min-100 mm/min welding speed. The two plates of AA6061 and AZ31B were butt welded to each other by FSW. AA6061 is placed at the advancing side and AZ31B is placed at the retreating side. The specimens for tensile test are prepared using ASTM E8. Figure 1 show ASTM E8 design. Then use electrical discharge machine (EDM). The tensile test carried out using Instron test machine using strain rate of 1 mm/min.

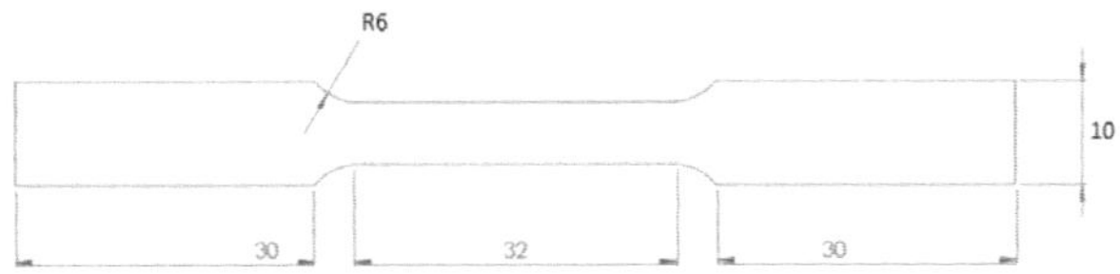

Figure 1. ASTM E8 design.

In the microstructural observation, etching is used to reveal the microstructure feature and remove dust or any small metal at the surface during polishing. The etching process was conducted by applying the etchants to the specimen and let it dry for few seconds. The process was conducted in fume hood to limit exposure of chemical liquid to air and dust. Microhardness of the specimen measured by using Vickers hardness using 300 kgf load and dwell time of 10s.

3. Results and Discussion

3.1. Weld Appearance

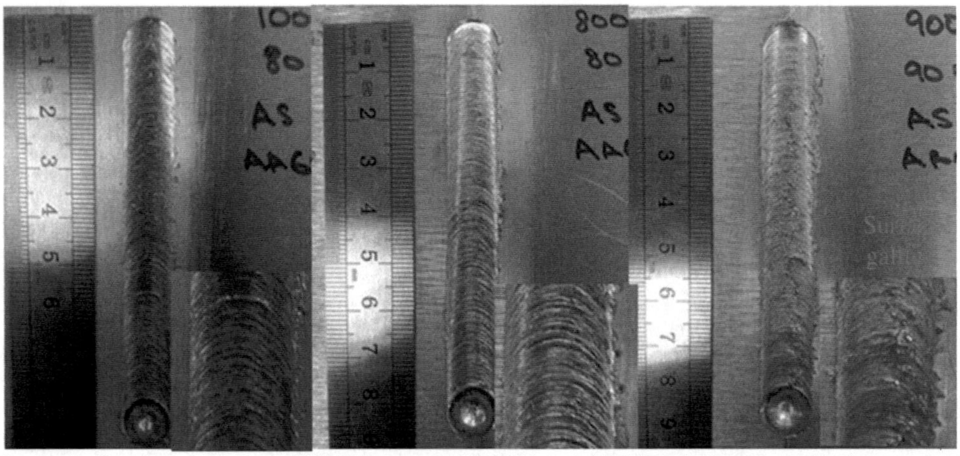

Figure 2. Weld appearance of FSW between AA6061 and AZ31B (a) Sample 2 (b) Sample 1 (c) Sample 9.

Figure 2 (a) is the sample 2 of FSW plate by using parameter 1000 rpm, 80 mm/min while Figure 2 (b) is the sample 1 of FSW plate by using 800 rpm and 80 mm/min. Both of these plates are free from defects. Sample 2 has a higher joint strength compared to sample 1 and sample 9 because it has enough heat during FSW process. Figure 2 (c) is the sample 9 of FSW plate by using 900 rpm and 90 mm/min. This plate is using the different heat input compare to sample 2 and sample 1 because of the different rotational speed and welding speed. Table 3 show the heat input of the plate of specimen 2, specimen 1 and specimen 9 during FSW process. The highest heat input is from the specimen 2 which is 0.4575KJ/mm. This show that specimen 2 receive enough heat input compare to specimen 9 which is affect the highest tensile strength. However, sample 9 shows the surface galling surface defect. It because it has less heat input from the sample 2 and inaccurate FSW parameter and affect from sticking metal from previous welding tool pin. Surface galling occur due to sticking of metal to the tool during welding causes the material on weld surface tearing up [10].

Table 3. Equation value for the power consumed and heat inputs for FSW [11].

Specimen	RPM	T	P	f_1	V
Specimen 1	800 rpm	1220 Nm	32533 W	0.9	80 mm/min
Specimen 2	1000 rpm	1220 Nm	40666 W	0.9	80 mm/min
Specimen 9	900 rpm	1220 Nm	36600 W	0.9	90 mm/min

3.2. Tensile Test

Table 4 show the tensile properties of the base materials and the weld specimen. The tensile result of the welded specimen is 76% of AZ31B. Figure 3 show the welded specimen failed at the centre joining weld and the friction stir welded specimen failed through brittle-mode fracture. Table 4 show the comparison of FSW with filler and without filler. FSW with filler is the highest tensile strength and joining efficiency which is 198 MPa and 76 % and FSW without filler is 175 MPa and 67 %. This result shows the joining efficiency increase from 67 % to 76 %. It is increase about 9 % of joining efficiency.

Table 4. Tensile test result of AA6061, AZ31B and FSW of AA6061 and AZ31B with no filler.

Material	Tensile strength (MPa)
Base metal AA6061	310
Base metal AZ31B	260
WELD	198
FSW of AA6061 and AZ31B with no filler	175

Table 5. Comparison of joining efficiency between types of FSW materials.

Type of FSW material	Joining efficiency
With filler	76 %
Without filler	67%

Figure 3 illustrates the tensile test data. The graph shows the Specimen 2 with the parameter of 1000 rpm, 80 mm/min and 3° of tilt angle has the highest tensile strength which is 198.216 MPa. Specimen 1 with the parameter of 800 rpm, 80 mm/min and 3° of tilt angle has the tensile strength of 165.616 MPa. The third highest tensile strength is 155.941 MPa was obtained by Specimen 4 with the parameter of 1000 rpm, 100 mm/min and 3° of tilt angle. Figure 4 show the tensile test comparisons between FSW welding with filler that is 198 MPa and without filler that is 175 MPa. It shows FSW with filler has higher tensile value.

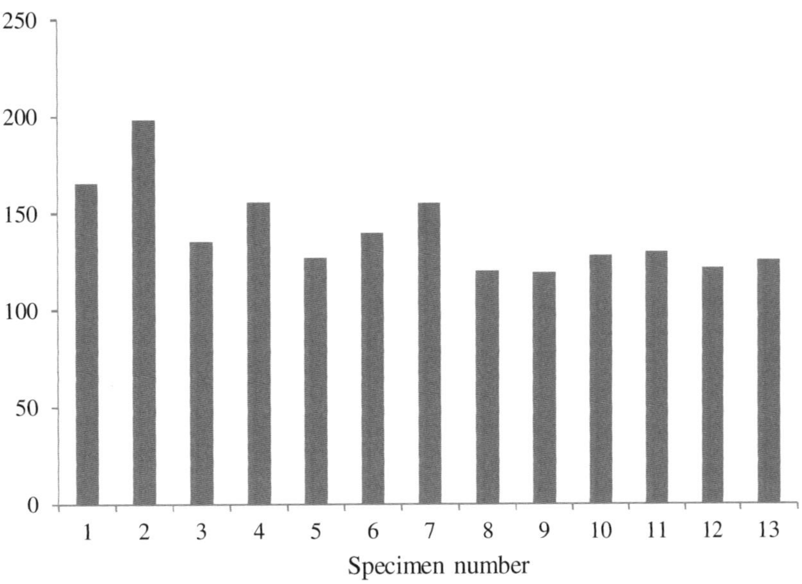

Figure 3. The tensile data.

Figure 5 shows the tensile fractured location. All joints fail at the stir zone of the weld. This is because stir zone at the center is the location which has the highest hardness value will cause the fractured brittle mode. Every dissimilar welding will occur fracture brittle mode [12]. During tensile testing, fracture occurred along the brittle and weak intermetallic layers. The resultant fracture surface shows the cleavage-type brittle fracture, as is shown in the SEM image in Figure 6 (b) [13]. This fracture is distinctly different from the microvoid type ductile fracture in the similar metal weld of 6061 Al in Figure 6 (a). It can be observed that the fractured surface was characterized with distribution of dissimilar size dimples [14]. These fine dimples indicate ductile behaviour of the aluminium alloy before the failure occurred.

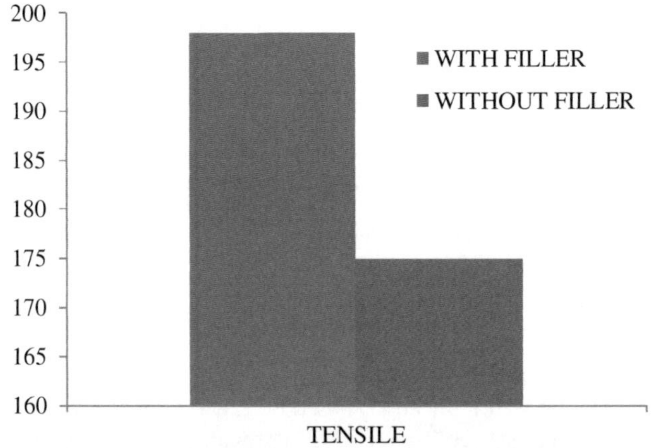

Figure 4. Tensile test comparisons between FSW welding with filler and without filler.

Figure 5. The tensile fractured location.

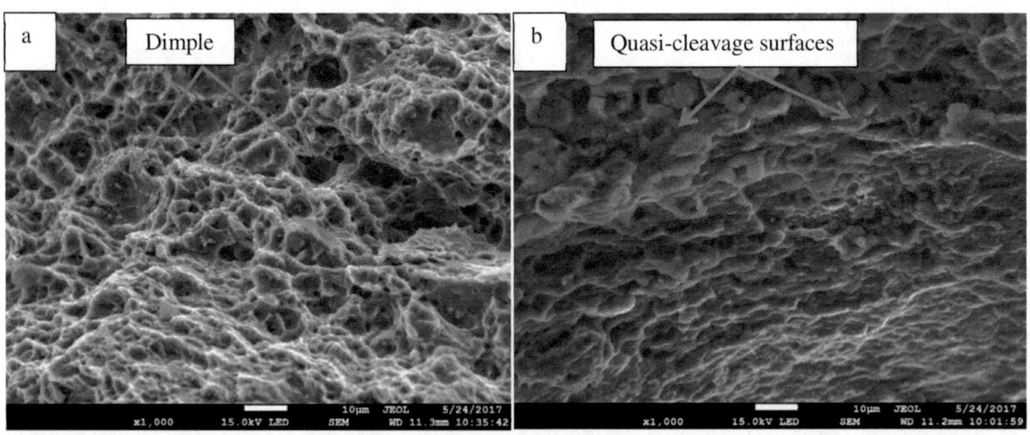

Figure 6. (a) SEM images of fracture surface on AA6061 (b) SEM images of fracture surface on AZ31B.

For the IMC layer observation in figure 7 (a) and figure 7 (b), it shows the FSW welding with filler have lower thickness of IMC layer than the FSW without filler IMC layer. The IMC layer for FSW without filler is 5.62 μm compare to IMC layer for FSW with filler that is 1.41 μm. The lower thickness of IMC layer will be resulting lower brittle of center of the welding. It also has lower hardness result for welding.

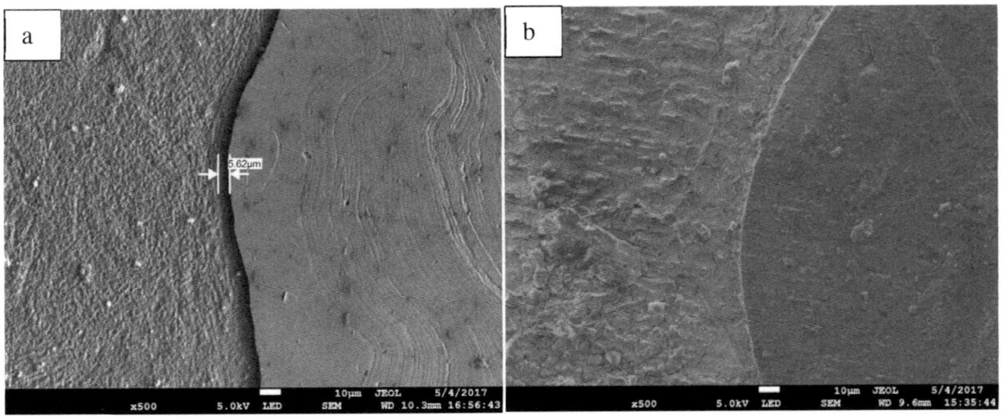

Figure 7. (a) SEM of IMC layer of FSW without filler (b) SEM of IMC layer of FSW with filler.

3.3. Microhardness

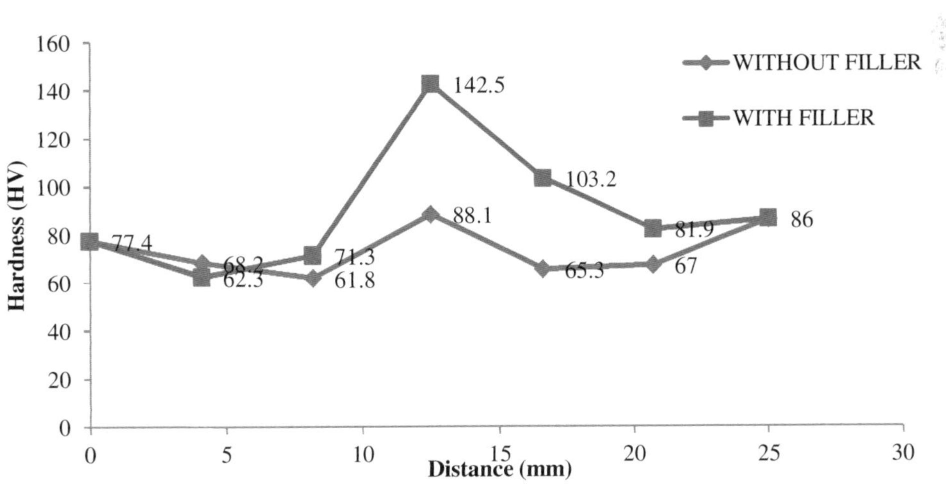

Figure 8. Vickers hardness comparisons between FSW with filler and without filler.

From Figure 8, it can be observed that the base metal (BM), heat affected zone (HAZ), thermo-mechanically affected zone (TMAZ) and stir zone (SZ) region can be differentiated by analysing the hardness value. It can be observed that AA6061 have a higher hardness value compared to AZ31B which is 86 HV compare to 77.4 HV. For both AA6061 and AZ31, it can also be identified that the hardness value at HAZ region will decrease from the hardness value of BM region. The lowest hardness value in this specimen can be identified at the HAZ region of AZ31B where the hardness value is 62.3 HV. The microhardness value at TMAZ was slightly higher than the HAZ region for both AA6061 and AA7075. The maximum hardness value of 142.5 HV has been achieved at the center of the weld. The maximum hardness value of 142.5 HV has been achieved at the center of the weld. Figure 4.23 below show the comparison of Vickers hardness between FSW using filler and without filler. With using filler, the result shown it have higher hardness value that is 142.5 HV compare to FSW without filler that is just 88.1 HV at the stir zone. From the lower result of hardness of welding without filler, it is because it has worm hole defect at the stir zone and higher grain size cause lower hardness. Figure 9 show the grain size of FSW without filler is higher than FSW with filler. It shows the entire zone at FSW without filler have higher grain size compare to FSW with filler. Then, higher grain size will affect lower hardness value. At the all zone using filler have higher hardness value except at the heat affected zone (HAZ) of AZ31B. The higher hardness value of FSW using filler resulting the higher tensile strength compare the FSW without filler.

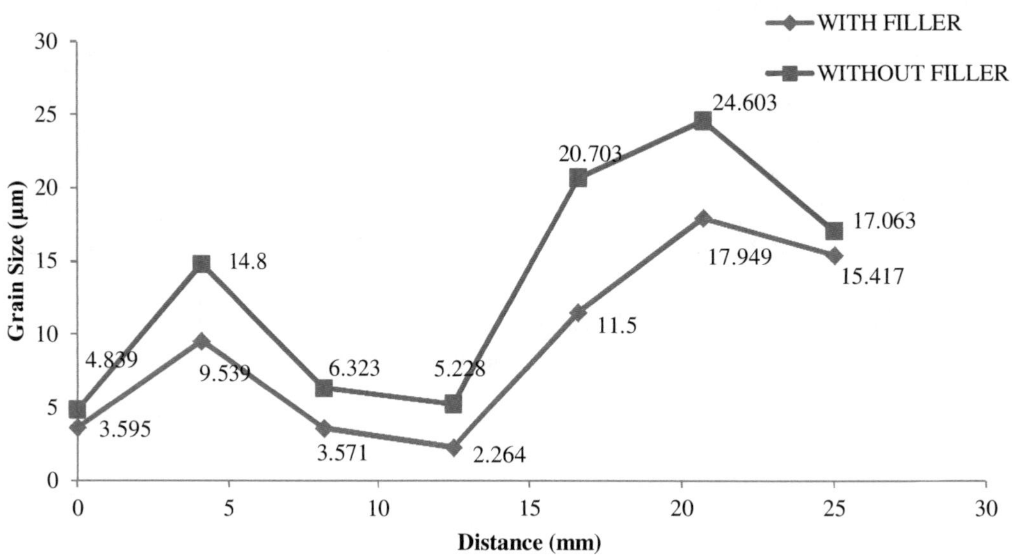

Figure 9. Grain size comparisons between FSW with filler and without filler.

4. Conclusion

- AZ31B magnesium alloy and AA6061 aluminium alloy with ER5356 filler can be join successfully by FSW.
- The optimum parameter for FSW between AA6061 and AZ31B using ER5356 which use response surface method (RSM) is when the rotational speed (RS) is 1000 rpm, welding speed (WS) is 80 mm/min and tilt angle is 3°. The tensile strength of the FSW by using the optimum parameter is 198 MPa and the highest hardness is 142.5 HV.
- The FSW microstructure change of welds using filler better than FSW without using filler because the grain is smaller and the IMC layer thickness is thinner which is 1.41 µm compare to IMC layer for FSW with filler that is 5.62 µm.
- The hardness profile shows that the center of the weld has the highest hardness value because brittle mode fractured.
- The joining efficiency of AZ31B magnesium alloy and AA6061 aluminium alloy use filler ER5356 is about 76% from lower base metal tensile strength. Then, the joining efficiency of AZ31B magnesium alloy and AA6061 aluminium alloy without using filler ER5356 is about 67%. This show FSW with filler is better about 9% increases of joining efficiency.

References

[1] Morishige T, Kawaguchi A, Tsujikawa M, Hino M, Hirata T and gashi K 2008 Dissimilar welding of Al and Mg alloys by FSW *Materials Transactions* **49** (5) 1129-1131

[2] Upadhyay P and Reynolds A P 2012 Effects of forge axis force and backing plate thermal diffusivity on FSW of AA6056 *Materials Science and Engineering: A* **558** 394-402
[3] Mishra R S and Ma Z 2005 Friction stir welding and processing *Materials Science and Engineering: R: Reports.* **50** (1) 1-78
[4] Gibson B Lammlein D H, Prater T J 2014 Friction stir welding: process, automation, and control *Journal of Manufacturing Processes* **16** (1) 56-73
[5] Li C, MuneharuaK, Takao S and Kouji H2009 Fiber laser-GMA hybrid welding of commercially pure titanium *Materials & Design* **30** (1) 109-114
[6] Commin L, Dumont M, Masse J E and Barrallier L 2009 Friction stir welding of AZ31 magnesium alloy rolled sheets: Influence of processing parameters *Acta materialia* **57** (2) 326-334
[7] Chen Y and Nakata K 2008 Friction stir lap joining aluminum and magnesium alloys *Scripta Materialia* **58** (6) 433-436
[8] Malarvizhi S and Balasubramanian V2012 Influences of tool shoulder diameter to plate thickness ratio (D/T) on stir zone formation and tensile properties of friction stir welded dissimilar joints of AA6061 aluminum–AZ31B magnesium alloys *Materials & Design* **40** 453-460
[9] Masoudian A 2014 Microstructure and mechanical properties of friction stir weld of dissimilar AZ31-O magnesium alloy to 6061-T6 aluminum alloy *Transactions of nonferrous metals society of China* **24** (5) 1317-1322
[10] Thompson B T 2010 Tool degradation characterization in the friction stir welding of hard metals. *The Ohio State University.*
[11] Lienert T, Stellwag W Jr and Lehman L 2002 Comparison of heat inputs: friction stir welding vs. arc welding *Columbus, OH, Edison Welding Institute*
[12] Yan J et al 2005 Microstructure characteristics and performance of dissimilar welds between magnesium alloy and aluminum formed by friction stirring *Scripta Materialia* **53** (5) 585-589
[13] Firouzdor V and Kou S 2010 Al-to-Mg friction stir welding: effect of material position, travel speed, and rotation speed *Metallurgical and Materials Transactions A* **41** (11) 2914-2935
[14] Xu W 2009 Temperature evolution, microstructure and mechanical properties of friction stir welded thick 2219-O aluminum alloy joints *Materials & Design* **30** (6) 1886-1893

IOP Conf. Series: Materials Science and Engineering **238** (2017) 012003 doi:10.1088/1757-899X/238/1/012003

The Effect of Tool Pin Shape of Friction Stir Welding (FSW) on Polypropylene

Z C Nik, M Ishak, N H Othman

Faculty of Mechanical Engineering, Universiti Malaysia Pahang, 26600, Pekan, Pahang, Malaysia

E-mail: zakwanzack1994@gmail.com

Abstract. This experiment deals with similar joining of polypropylene (PP) with thickness of 3 mm was carried out by using friction stir welding (FSW) technique. The process parameters, rotational speed, welding speed and tilt angle were fixed of experiments. The tool geometry shapes were the main parameters which were taken into consideration. The optimum designs of tool geometry shape were determined with reference to tensile strength of the joint. During the tensile testing experiment, the results show that all fractured occurs in the heat-affected zone (HAZ) on the polypropylene (PP). Results show that the optimum design can be obtained with same rotational speed, welding speed and tilt angle.

1. Introduction

Magnesium Friction Stir Welding (FSW) is a welding method under solid state joining that involves no melting that is welding below melting point [1]. The advantages of FSW compared to another method are excellent mechanical properties in fatigue, tensile and bend test, no arc or fume, can weld in all position, no filler wire needed, weight reduction because no usage of filler and uses consumable tools which means that the tool can used for many times [2]. FSW tool consists of two parts, namely shoulder and pin. The functions of pin are the primary source for material deformation and the secondary source for heat generation in the nugget [3]. The pin is designed to disrupt the faying or contacting surface of the work piece, shear material in front of the tool and move material behind the tool [4]. Shoulder is the main source of heat generated during the process, the primary constraint to material expulsion and primary driver for material flow around the tool [3]. Tool shoulder is designed to produce heat to the surface and subsurface regions of the welding material [4]. They are several types of tool profile such as cylindrical, threaded pin and taper pin profile. However, each uses different uses different pin tool shape geometry and it affects the strength, macrostructure and microstructure of the specimen. This study investigates the effect of tool pin shape of friction stir welding (FSW) on polypropylene.

2. Experimental Procedure

2.1. Material

The base material used in this study was polypropylene (PP) with 3 mm thickness. The dimension of base metal to be welded was 60 mm (width) × 100 mm (length) × 6 mm (thickness). The plate was longitudinally butt welded using PARTNER VMM3917 conventional milling machine. Polypropylene is a linear hydrocarbon polymer, $(C_3H_6)_n$. Polypropylene is a gaseous compound obtained by the thermal cracking of ethane, propane, butane and the naphtha fraction of petroleum. Table 1 shows the mechanical properties of polypropylene (PP).

Table 1. Mechanical properties of polypropylene (PP) material.

Material	Density	Yield Strength	Melting Point
Polypropylene	0.855g/cm³	12.43 MPa	130-171°C

2.2. Welding Setup

The process parameters used for this experiment are given in Table 2 and the parameters are set from a defect free specimens. These parameters; welding speed, rotational speed and tilt angle were kept constant throughout the process. The welding tool material used is H13 steel. The tool pin shapes used in this study were cylindrical tapered, hexagon and triangle. Detailed tool specifications are presented in Table 3.

Table 2. Welding parameters for FSW.

Parameters	Values
Rotational speed	1500 RPM
Welding speed	45 mm/min
Tilt angle	1°

Table 3. FSW tool specifications.

Items	Values (mm)
L, length of pin	2.80
d, diameter of pin	3.00
D, diameter of shoulder	10.50

Figure 1. (a) Cylindrical tapered (b) Hexagon (c) Triangle.

Tensile test was performed to evaluate the mechanical properties of the joints. The tensile test was prepared according to the ASTM B 557 M – 07 standards. The tensile test was conducted with one sample have three repetitions and the average values were presented in result and discussion. The data from the tensile test was recorded and analyzed accordingly. A metallographic specimen was performed to analyze the microstructure of welding area. To analyze the microstructure the cross-sectional specimens were then cut and mounted. The mounted samples were grinding manually using different grades sand papers; 240, 320, 600. 2400. The microstructure of the weld specimens was observed using an optical microscope.

3. Results and Discussion

3.1. Weld Appearance

The FSW weld appearance of polypropylene (PP) of different pin tool geometry shown in Figure 2. The inset in each sample shows a closer view of the weld scan. Figure 2(a) is the sample of FSW plate by using cylindrical tool profile with parameter 1500 rpm, 45 mm/min and 1° of tilt angle while Figure 2(b) is the sample of FSW plate by using hexagon tool profile with same parameter. The Figure 2(c) shows the sample using triangle tool profile also using with same parameter. All of the plates that show on that figures are defect-free samples. It also worth noting that from the tensile test conducted below, sample (a) has a higher joint strength compared to sample (b) and sample (c).

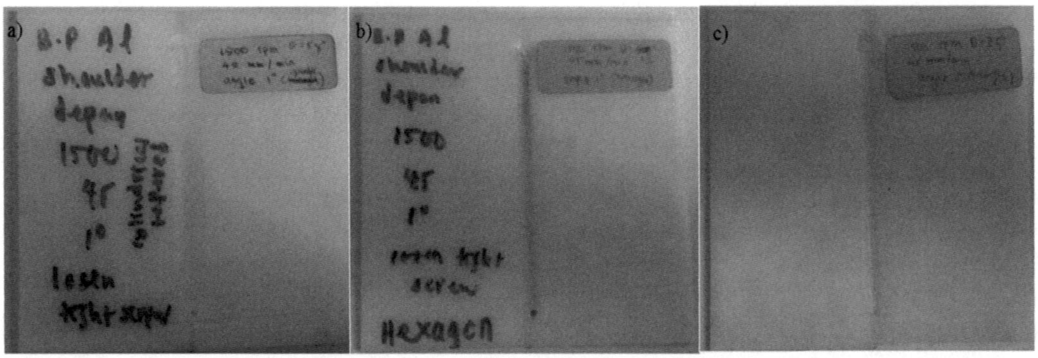

Figure 2. (a) Weld appearance cylindrical tapered pin shape (b) Weld appearance hexagon pin shape (c) Weld appearance triangle pin shape.

3.2. Tensile Test

Table 4 illustrates the tensile test data in table. The graph shows the sample number 2 for cylindrical using cylindrical tapered tool shape with parameter 1500 rpm, 45 mm/min and 1° of tilt angle has the highest tensile strength which is 8.95 MPa. The sample number 2 for triangle using triangle tool shape with same parameter has the tensile strength of 5.78 MPa. The third highest tensile strength is 5.62 MPa was obtained by sample number 1 for hexagon using hexagon tool profile with same parameter that are 1500 rpm, 45mm/min and 1° tilt angle.

Table 4. Result tensile test.

Number of samples	Ultimate tensile strength (MPa)	Tool geometry
C1	6.57	Cylindrical tapered
C2	8.95	Cylindrical tapered
C3	8.63	Cylindrical tapered
H1	4.73	Hexagon
H2	5.62	Hexagon
H3	1.50	Hexagon
T1	5.29	Triangle
T2	5.78	Triangle
T3	2.46	Triangle
Base metal	23.0	

Figure 3 show that the tensile fractured location. This test is done to find where the breakage occurs in the rear side of the welded polypropylene. This test concludes that there was lower strength on the back side of the weld, so the breakage was initiated from the point on the rear side of the welded specimen. According to experiment that conducted for the base metal tensile strength, the ultimate tensile strength for polypropylene (PP) is 23 MPa. The welding efficiency shows from equation for cylindrical tapered pin shape tool profile is the strongest, 35% compare to hexagon and triangle pin shape tool profile. The welding efficiency for hexagon and triangle pin shape tool profile is 23% and 24% respectively. On the other hand, hexagon pin shape tool profile is the weakest of tensile strength. The cylindrical tapered pin shape tool is the highest welding efficiency because the composition of polypropylene (PP) using this tool is concentrated may cause an inclusion defects in weld cross section [5]. After that, composition of polypropylene (PP) become low and may causes porosity in weld cross section [6]. Because of that, hexagon pin shape tool is the lowest welding efficiency.

Figure 3. Tensile test fracture of polypropylene (PP).

3.3. Angular Distortion

Distortion is any unwanted physical change or departure from specifications in a fabricated structure or components, as a consequence of welding. In this experiment angular distortion was appeared. Angular distortion is weld tends to be wider at the top than the bottom causing more solidification shrinkage and thermal contraction [7]. The angular distortion occurs after the first run of weld cools, contracts and draws the plates together. The second run has the same shrinkage effect but its contraction is restricted by the solidified first run which acts as fulcrum for angular distortion.

Cylindrical tapered is the highest of angular distortion occurred, 5.8°. After that, angular distortion occurred in 3.9° using triangle pin shape tool. Then, the lowest is hexagon pin shape tool that angular distortion occurred in 2.2°. The highest angular distortion occurred at cylindrical tapered because the pin profile is flat faces and associated with eccentricity. This eccentricity allows incompressible material to pass around the pin profile [8]. Because of that, the highest an angular distortion obtained using cylindrical tapered pin shape tool. After that, an angular distortion is lowest when using hexagon and triangle pin shape tool cause that profiles produce an energetic stirring action in the flowing material due to flat surfaces [8]. From the data, it can be concluding that the angular distortion effects the tensile strength. When the angular distortion is increase the tensile strength also increase.

3.4. Macrostructure analysis

In this experiment the polypropylene (PP) material through the process of Friction Stir Welding (FSW). The composite of polypropylene (PP) material is stir mix together to form a weld joint and the material can be joint. Figure 6 shows the reason why cylindrical tapered pin shape tool is the highest tensile strength. In cylindrical tapered the inclusion defect is appeared at cross section weld line. The inclusion occurs make the weld joint is become stronger and the tensile strength is the highest [9] compare to hexagon and triangle pin shape tool. Inclusion occurs because when the pin shape tool is plunge into polypropylene (PP) material the stir process between pin shape tool and polypropylene (PP) occurs and may cause the composite of polypropylene (PP) is concentrated. When the composite polypropylene (PP) concentrated, it was solidified at weld joint. The solidification of concentrated composite polypropylene (PP) called inclusion defect. After that, from Figure 7 the wormhole defect is appeared. The wormhole occurs at cross section weld line of hexagon pin shape tool. The effect of wormhole may affect the tensile strength of weld line [10]. These defect causes the hexagon pin shape tool is the lowest tensile strength. Wormhole occurs because the composite of polypropylene (PP) material is concentrated during stir when the Friction Stir Welding process occurred. Because of that wormhole defect occurs at the cross section of weld line on polypropylene material when using hexagon pin shape tool. On the other hand, the porosity is obtained in Figure 4. Porosity appeared using triangle pin shape tool. It is because insufficient of heat generated from the friction of tool and it not stir fully well during welding process. The effect of porosity causes the tensile strength become low.

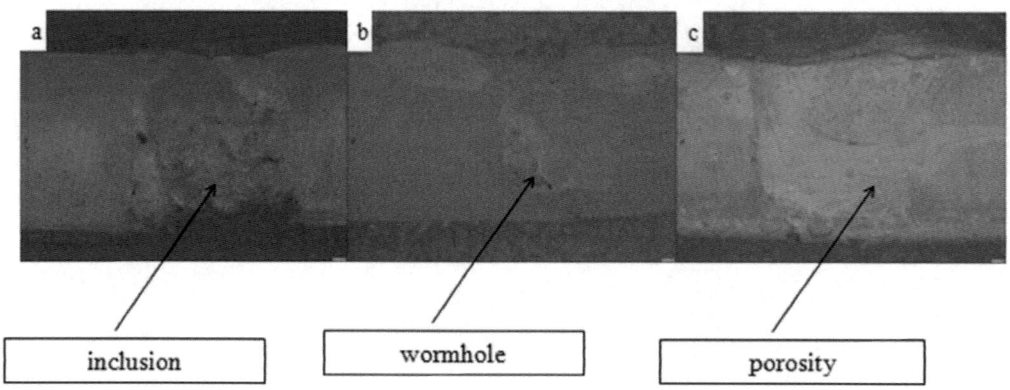

Figure 4. (a) Weld cross section cylindrical tapered pin shape tool (b) Weld cross section hexagon pin shape tool (c) weld cross section triangle pin shape tool.

3.5. Microstructure analysis

In Figure 5 the microstructure using cylindrical tapered pin shape tool is showed. From that figure the tunnel defect is obtained. When the tunnel defect is obtained the angular distortion was formed because the deformation between polypropylene (PP) materials is occurred. The tunnel defect is occurred at weld line. Because of that it when using cylindrical tapered pin shape tool, it will more deform compare triangle and hexagon pin shape tool. The residual stress between these tools, cylindrical tapered pin shape tool is more residual stress.

In hexagon pin shape tool, the microstructure is same as like cylindrical pin shape tool. It showed in Figure 5. But, it a little bit different compare cylindrical tapered it is a porosity defect is appeared. The effect of porosity will affect tensile strength, the tensile strength become low and also the angular distortion is less compare to triangle and hexagon pin shape tool [11]. This reason hexagon pin shape tool is lowest tensile strength. After that, for triangle pin shapes tool it also same to cylindrical tapered and hexagon type pin shape tool. But in microstructure triangle pin shape tool the area of tunnel defect is smaller to cylindrical tapered pin shape tool. It had been showed in Figure 5. It is because triangle pin shape tool is less residual stress compare to cylindrical tapered pin shape tool. Inclusion and porosity are also appeared in triangle pin shape tool. Inclusion make the tensile is high compare to hexagon but cylindrical tapered pin shape tool is the highest. In conclusion, the highest angular distortion using cylindrical tapered pin shape tools while the lowest angular distortion using hexagon pin shape tool.

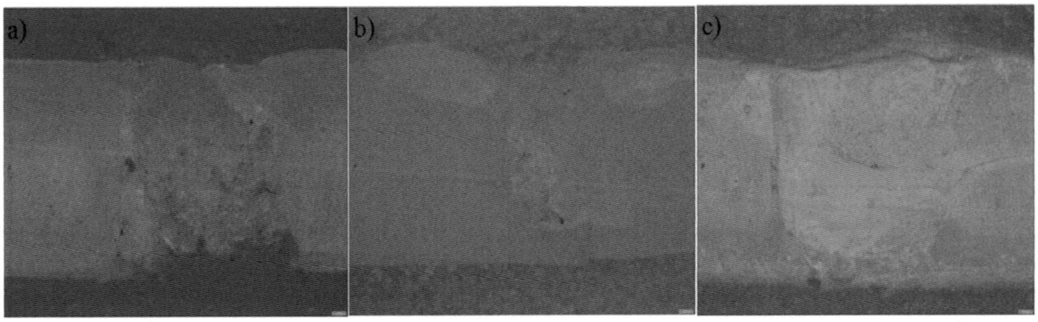

Figure 5. (a) Microstructure cylindrical tapered pin shape tool (b) Microstructure hexagon pin shape tool (c) Microstructure triangle pin shape tool.

4. Conclusion

The fabrication of joining 3 mm thick polypropylene (PP) plates were successfully done by using Friction Stir Welding (FSW) process with fix parameter of welding speed, rotational speed and tilt angle. It is also done by using different tool geometry and was butt joint. The highest tensile strength is 8.95 MPa using a cylindrical pin shape tool compare hexagon and triangle pin shape tool with tensile strength values of 5.62 MPa and 5.78 MPa respectively. The macrostructure analysis shows for cylindrical tapered pin shape tool an inclusion was obtained while hexagon pin shape tool wormhole defect was occurred and triangle pin shape tool porosity was observed. The microstructure analysis shows that polypropylene (PP) will produce tunnel defect at the welding cross section. This tunnel defect caused an angular distortion obtained for polypropylene (PP) material. The optimum for pin tool geometry in this research is cylindrical tapered pin shape tool because it is the highest tensile strength value.AZ31B magnesium alloy and AA6061 aluminium alloy without using filler ER5356 is about 67%. This show FSW with filler is better about 9% increases of joining efficiency.

References

[1] Siddiqui M A et al 2014 Friction Stir Welding as a Joining Process through Modified Conventional Milling Machine: A Review *International Journal of Innovative Research and Development* ISSN **3** (7) 2278–0211

[2] Baghel P K 2012 Design and development of fixture for friction stir welding 172-175

[3] Meilinger Á and Török I 2013 The importance of friction stir welding tool *Production Processes and Systems* 2013. **6** (1) 25-34

[4] Eslami S et al 2015 Effect of friction stir welding parameters with newly developed tool for lap joint of dissimilar polymers *Procedia Engineering* **114** 199-207

[5] Mehta K P and Badheka V J 2016 Effects of Tool Pin Design on Formation of Defects in Dissimilar Friction Stir Welding *Procedia Technology* **23** 513-518

[6] Jaiganesh V Maruthu B and Gopinath E 2014 Optimization of process parameters on friction stir welding of high density polypropylene plate *Procedia Engineering* **97** 1957-1965

[7] Sadeesh P et al 2014 Studies on Friction Stir Welding of AA 2024 and AA 6061 Dissimilar Metals *Procedia Engineering* **75** 145-149

[8] Bilici M K and Yükler A I 2012 Influence of tool geometry and process parameters on macrostructure and static strength in friction stir spot welded polyethylene sheets *Materials & Design* **33** 145-152

[9] Banjare P N Sahlot P and Arora A 2017 An assisted heating tool design for FSW of thermoplastics *Journal of Materials Processing Technology* **239** 83-91
[10] Mendes N et al 2014 Effect of friction stir welding parameters on morphology and strength of acrylonitrile butadiene styrene plate welds *Materials & Design* **58** 457-464
[11] Rezaee H M et al 2017 Investigation on the effects of tool geometry on the microstructure and the mechanical properties of dissimilar friction stir welded polyethylene and polypropylene sheets *Journal of Manufacturing Processes* **26** 269-279

Resistance Element Welding of Magnesium Alloy/austenitic Stainless Steel

S M Manladan [1,3], F Yusof [1,2*], S Ramesh [1,2], Y Zhang [4], Z Luo [4,5], and Z Ling [4]

[1] Department of Mechanical Engineering, Faculty of Engineering, University of Malaya, 50603 Kuala Lumpur, Malaysia
[2] Center for Advanced Manufacturing and Materials Processing (AMMP), Faculty of Engineering, University of Malaya, 50603 Kuala Lumpur, Malaysia
[3] Department of Mechanical Engineering, Faculty of Engineering, Bayero University, Kano, 3011 Kano, Nigeria
[4] School of Materials Science and Engineering, Tianjin University, Tianjin 300072, China
[5] Collaborative Innovation Center of Advanced Ship and Deep-Sea Exploration, Shanghai 200240, China

E-mail: farazila@um.edu.my

Abstract. Multi-material design is increasingly applied in the automotive and aerospace industries to reduce weight, improve crash-worthiness, and reduce environmental pollution. In the present study, a novel variant of resistance spot welding technique, known as resistance element welding was used to join AZ31 Mg alloy to 316 L austenitic stainless steel. The microstructure and mechanical properties of the joints were evaluated. It was found that the nugget consisted of two zones, including a peripheral fusion zone on the stainless steel side and the main fusion zone. The tensile shear properties of the joints are superior to those obtained by traditional resistance spot welding.

1. Introduction

Multi-material design is an efficient way to reduce the weight of vehicle structures, improve crash-worthiness, balance cost, and reduce environmental pollution [1]. Magnesium alloys and stainless steels have been identified as excellent materials for the next generation vehicles. Mg alloys possess remarkable properties such as superior specific strength, high elastic modulus, hot formability, good cast ability, and recycle ability [2, 3]. On the other hand, austenitic stainless steels (ASS) have excellent combination of corrosion resistance, decorative appearance and mechanical properties, such as high strength, unique work-hardening behavior, superior energy absorption ability, and high formability. They are commonly used in construction of the body frame of buses and passenger rail vehicles [4]. Furthermore, the next generation vehicle project shows that ASSs can replace carbon steels in crash-relevant components, thereby reducing weight without compromising vehicle crash-worthiness [5].

With the growing interest in multi-material design and the fact that both Mg alloy and ASSs are increasingly attracting attention in the transportation industry, it is imperative to develop an efficient technique to join them in dissimilar combination. The main sheet joining technique in the transportation industry is resistance spot welding (RSW) because it is inexpensive, easy to operate and automate, fast and reliable, etc. [6, 7]. However, it is difficult to join Mg alloy to steels by RSW because of large differences in physical and metallurgical properties [8]. A variant of RSW, known as resistance element welding (REW) was recently developed for joining dissimilar steel/light alloys combinations that are metallurgical incompatible and difficult to weld together [9]. The techniques involve punching a technological in the light alloy and inserting a steel rivet (an auxiliary element) into the hole, followed by RSW on the rivet/steel [10, 11]. So far, it has been used to join Al alloys to steels [10, 11] and AlSi-coated 22MnB5 steel to LITECOR® [12]. The mechanical performance of the joints obtained using this technique was found to be superior to those obtained by traditional RSW [10, 11].

This technique has not yet been applied to join Mg alloy to steel. Owing to the industrial importance of Mg alloy and ASS, the present study investigates the microstructure and mechanical properties of dissimilar Mg alloy/ASS joints. Traditional RSW was also conducted to compare the mechanical performance of the joints.

2. Experimental Procedure

The materials used in this study were 1.5 mm thick AZ31 Mg alloy and 0.7 mm thick AISI 316L ASS. Both materials were cut into rectangular specimens (100 x 25 mm) in accordance with AWS D17.2 Standard [12], and alcohol was used to clean all the specimens. In addition, abrasive paper was used to grind the surfaces of the Mg alloy specimens in order to remove surface oxides. A hole with diameter of 5 mm was punched at the center of the overlap area of the Mg alloy, and Q235 steel rivet was inserted into the hole as shown in Figure 1.

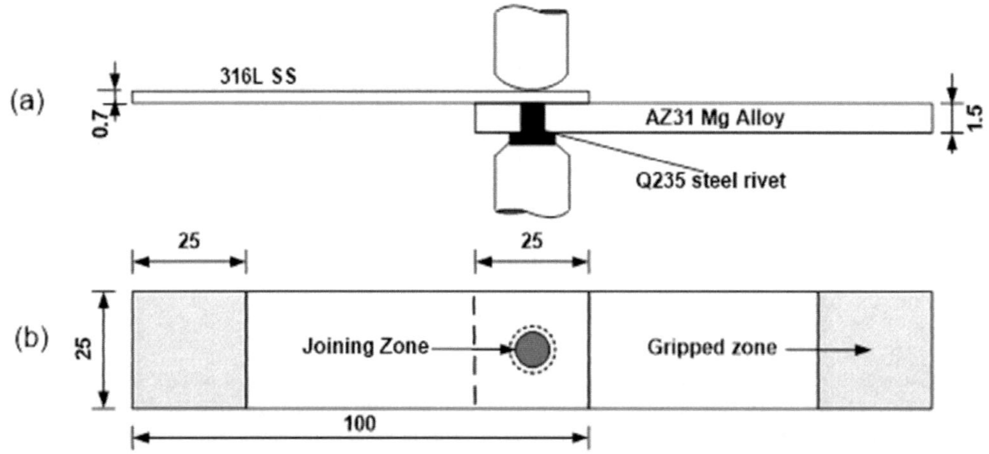

Figure 1. Schematic illustration of (a) REW process and (b) tensile shear test specimen.

The welding process was conducted using a 220 kW, medium-frequency direct current RSW machine with a capacity of 2–22 kA welding current. Asymmetrical electrodes were used with the aim of improving the heat balance [13]. A spherical electrode (50 mm sphere radius, 20 mm face diameter) was used on the Mg alloy side and a truncated cone electrode (10 mm tip diameter) was used against the rivet. Both electrodes were made of copper alloy (RWMA class II). The welding current was varied from 5-9 kA in 1 kA increment while the welding time and electrode force were kept constant at 250 ms and 3.6 kN, respectively.

For metallographic studies, samples were cut at the center of the joints, ground and polished in accordance with standard metallographic procedures. The Mg side was etched with a solution of 5 g picric acid, 10 ml H_2O, 5 ml acetic acid, and 100 ml ethanol. A solution 4% nital was used to etch the Q235 steel side while a solution of 10 g $FeCl_3$, 30 ml HCl, and 120 ml H_2O was used to etch the ASS side. The macrostructural and microstructural observations were conducted with an Olympus SZX12 stereomicroscope and Olympus GX51 microscope, respectively. Hardness tests were conducted under a load of 200 g and a dwell time of 15 s using Huayin HV-1000A Vickers micro-hardness tester. A CSS-44100 material testing system was used to conduct tensile shear tests at a cross-head speed of 2 mm/min. The samples configurations are shown in Figure 1. Three samples were tested for each welding condition and the average was evaluated. Fractography was conducted using a Hitachi SU1510 scanning electron microscope (SEM).

3. Results and Discussion

3.1. Macrostructures and microstructures

The typical macrostructure of the REW joint is shown in Figure 2 (a), and it can be seen that an asymmetrical nugget was obtained. The final solidification line and the larger part of the nugget are in the Q235 steel rivet. The formation of asymmetrical nugget can be attributed to the differences in electrical resistivity and thermal conductivity. The nugget can be divided in to two zones, namely, peripheral FZ (FZ1) and main FZ (FZ2), as shown in Figure 2 (b). The peripheral FZ, as shown in the figure, formed only on the ASS side. Two-zone nugget formation during RSW involving ASSs was attributed to the differences in volume fraction of delta ferrite between the main nugget and the periphery [14]. The solidification mode in ASS RSW occurs in ferrite-austenite (FA) mode, as follows [14]:

$$L \xrightarrow{I} L + \delta \xrightarrow{II} L + \delta + \gamma \xrightarrow{III} \delta + \gamma$$

The microstructure of the base metal (BM), which is austenitic, transforms in to austenite and intercrystalline delta ferrite. As a result of higher cooling rate at the periphery because of its closeness to the water-cooled electrodes, there is limited time for stage III reaction to occur. Therefore, the delta ferrite volume fraction is higher in the periphery.

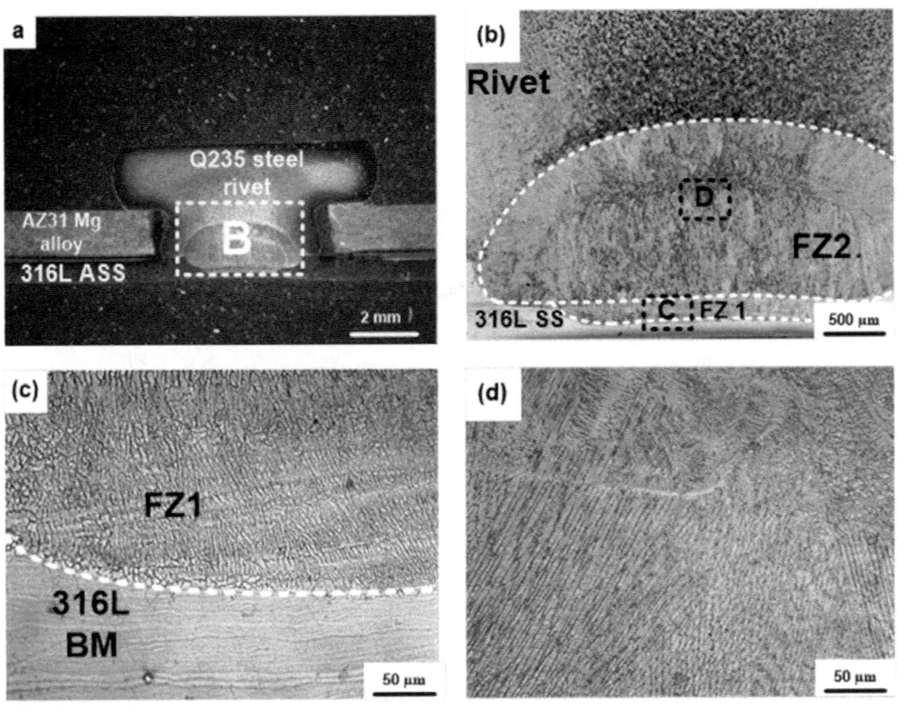

Figure 2. Macrostructure and microstructure of REW joint (a) macrostructure, (b) higher magnification of region B in (a), (C) microstructure of region C in (b), and (d) microstructure of region D in (b).

Figure 3 shows the microstructures in the heat affected zone (HAZ). As shown in Figure 3 (b), the HAZ can be divided in two different zones, namely; (i) the upper-critical HAZ (UCHAZ): the maximum temperature that is attained in this zone is above A_{C3}, and the BM completely transforms in to austenite. As a result of the fast cooling rate inherent in RSW, the austenite transforms in to martensite after cooling. (ii) inter-critical HAZ (ICHAZ): the peak temperature reached in this zone is between A_{C1} and A_{C3}, and the BM microstructure transformed into a mixture of ferrite and austenite. After cooling, the austenite transformed in to pearlite while the ferrite is retained. Thus, the microstructure consisted of a mixture of ferrite and pearlite (Figure 3 (c)). The BM microstructure also consisted of ferrite and pearlite (Figure 3 (d)), but the volume fraction of the pearlite is less than that of the ICHAZ.

As shown in Figure 3 (d), the microstructure of the BM can be divided in to BM1 (rivet shank) and BM2 (rivet cap). Both consisted of ferrite and pearlite. However, BM2 grains are deformed because of coldworking.

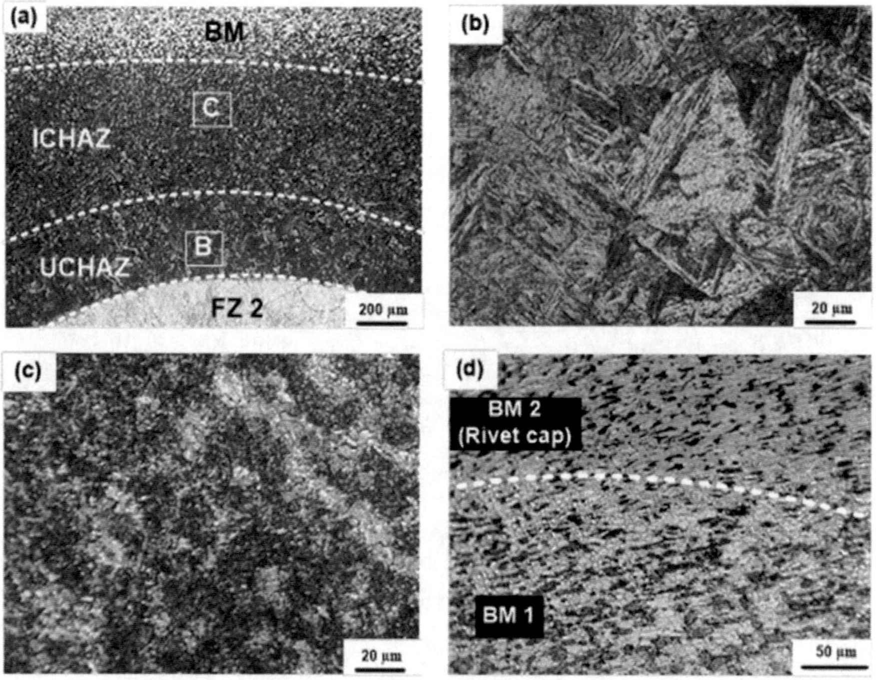

Figure 3. HAZ and BM microstructure of the Q235 steel rivet: (a) microstructural gradient, (b) higher magnification of region B in (a), (c) higher magnification of region C in (a), and (d) BM microstructure.

3.2. Hardness characteristics

The typical hardness profile of the REW joint is shown in Figure 4, indicating microstructural gradient across the weldment. The average hardness value of FZ2 is 401.3 HV, and that of FZ1 is 331.5HV. The high hardness value of FZ2 has been associated with some martensitic transformation [15]. An average hardness value of 427 HV, 282.3 HV, and 244.3 HV is obtained in the UCHAZ, ICHAZ, and BM respectively. The highest hardness value of 448 HV is obtained in the UCHAZ, close to FZ2. The average hardness of the BM2 is 331.5 HV and that of BM 1 is 244.3 HV. An average hardness value of 197 HV is obtained for ASS HAZ, which is lower than the hardness of the BM (212 HV). This hardness reduction is attributable to the loss of work hardening in the BM.

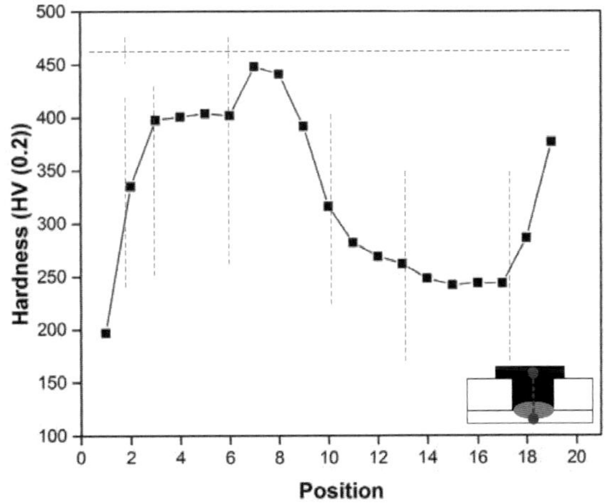

Figure 4. Hardness profile of REW joint.

3.3. Tensile-shear performance

Figure 5 shows the influence of welding current on the nugget diameter, energy absorption and peak load of the joints produced by REW. It can be seen that as the welding current is increased from 5 to 8 kA, the peak load and energy absorption increased from 1.85 kN and 2.85 J to 3.71 kN and 10.19 J, respectively. This increase in peak load and energy absorption is as a result of increase in nugget diameter. However, when the welding current was increased to 9 kA, the peak load and energy absorption dropped significantly despite the increase in nugget diameter. This is because the Mg alloy around the rivet melted excessively at this welding current, leading to the widening of the rivet hole.

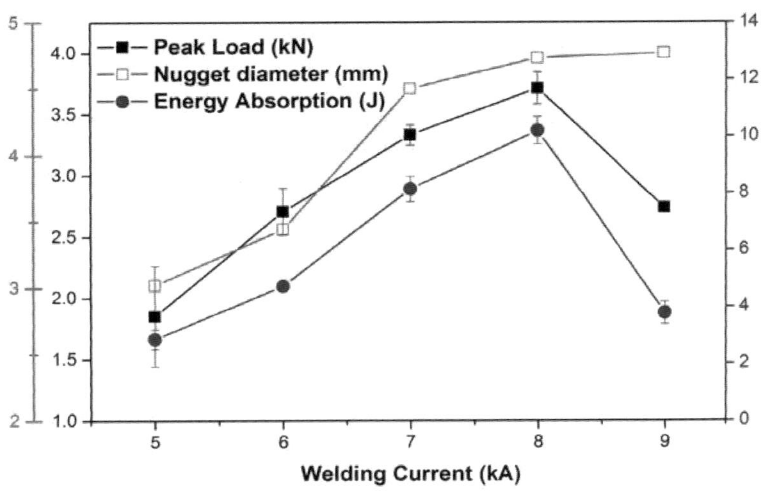

Figure 5. Effect of welding current on the nugget diameter, peak load and energy absorption of joints produced by REW.

For the purpose of comparison, joints were also produced using traditional RSW. The peak load obtained by REW is 3.71 kN and that obtained by RSW is 2.3 kN. The energy absorption for REW is significantly higher, which is very important for passenger safety in the event of a crash. The maximum energy absorption obtained by REW is 10.2J while that obtained by RSW is 1.14 J. In addition, the REW joints were obtained at lower welding currents. The peak load and maximum energy absorption of the REW were obtained at a welding current of 8 kA while those of RSW were obtained at 14 kA.

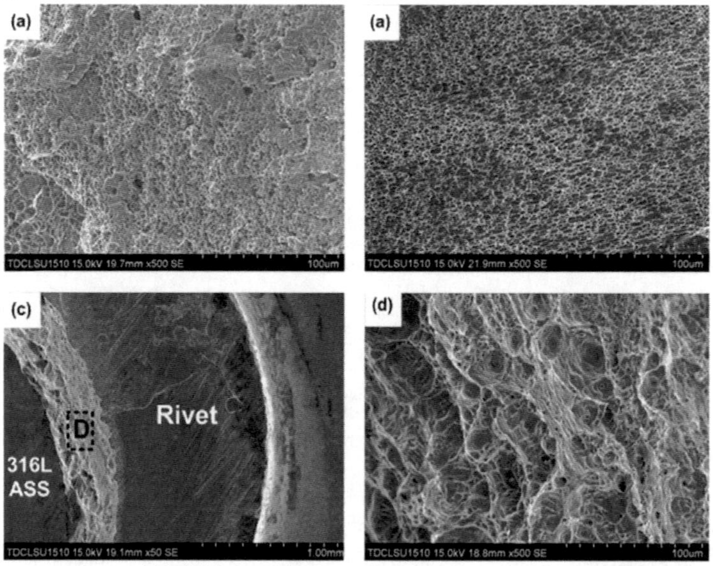

Figure 6. Fracture surface of REW joint (a) IF mode (Q235 steel side), (b) IF mode (ASS side), (c) macroscopic morphology of PO mode, and (d) microstructure of region D in (c).

The REW joints failed in interfacial failure (IF) mode at a welding current of 5 kA. With further increase in welding current, pullout failure (PO) mode, which has higher energy absorption, occurred. Figure 6 (a) and (b) show the fracture surfaces of the joints that failed in IF mode, which has a quasi-cleavage (combined ductile and brittle) characteristics. The PO mode fracture surface (Figure 6 (d)) has larger and deeper dimples, which indicate a ductile fracture.

4. Conclusion

The present study investigated the microstructure and mechanical properties of dissimilar joints between 1.5 mm AZ31 Mg alloy and 0.7 mm 316L ASS produced by REW processes. Traditional RSW joints were also produced to compare the mechanical performance. The following conclusions can be drawn:

- A two-zone nugget, consisting of a peripheral FZ on the stainless steel side and the main nugget at the rivet/stainless steel interface, was formed. Both the peripheral FZ and the main FZ consisted of austenite and inter crystalline delta ferrite.
- The mechanical performance of the REW joints is superior to that of traditional RSW joints.
- The failure mode of the REW joints changed from IF to PO with increase in welding current.

Acknowledgement

The authors would like to acknowledge University of Malaya, Malaysia and Tianjin University, China for providing the resources and facilities for this research. This research is funded by the Postgraduate Research Fund (PPP) with Grant No: PG020-2015A and University Malaya Research Grant (UMRG) with Grant No.: RP035A-15AET.

References

[1] Haddadi F 2016 Microstructure reaction control of dissimilar automotive aluminium to galvanized steel sheets ultrasonic spot welding *Materials Science and Engineering: A* **678** 72-84

IOP Conf. Series: Materials Science and Engineering **238** (2017) 012004 doi:10.1088/1757-899X/238/1/012004

[2] Babu N K Brauser S Rethmeier M and Cross C 2012 Characterization of microstructure and deformation behaviour of resistance spot welded AZ31 magnesium alloy *Materials Science and Engineering: A* **549** 149-156
[3] Manladan S Yusof F Ramesh S and Fadzil M 2016 A review on resistance spot welding of magnesium alloys *The International Journal of Advanced Manufacturing Technology* **865** 1805-1825
[4] Pouranvari M Khorramifar M and Marashi P 2016 Ferritic–austenitic stainless steels dissimilar resistance spot welds: metallurgical and failure characteristics *Science and Technology of Welding & joining*
[5] Schuberth S Schedin E Frohlich T and Ratte E 2008 Next generation vehicle engineering guidelines for stainless steel in automotive applications *6th Stainless Steel Science and Market Conference* Helsinki, Finland: The Swedish Steel Producers' Association
[6] Behravesh S B Jahed H and Lambert S 2011 Characterization of magnesium spot welds under tensile and cyclic loadings *Materials & Design* **3210** 4890-4900
[7] Manladan S Yusof F Ramesh S Fadzil M Luo Z and Ao S 2017 A review on resistance spot welding of aluminum alloys *The International Journal of Advanced Manufacturing Technology* **901** 605–634
[8] Liu L Xiao L Chen D Feng J Kim S and Zhou Y 2013 Microstructure and fatigue properties of Mg-to-steel dissimilar resistance spot welds *Materials & Design* **45** 336-342
[9] Holtschke N and Jüttner S 2016 Joining lightweight components by short-time resistance spot welding *Welding in the World* 1-9
[10] Ling Z Li Y Luo Z Feng Y and Wang Z 2016 Resistance Element Welding of 6061 Aluminum Alloy to Uncoated 22MnMoB Boron Steel *Materials and Manufacturing Processes* just-accepted
[11] Qiu R Wang N Shi H Cui L Hou L and Zhang K 2015 Joining steel to aluminum alloy by resistance spot welding with a rivet *International Journal of Materials Research* **1061** 60-65
[12] AWS 2007 Specification for resistance welding for aerospace applications D17.2 *American Welding Society*
[13] Liu L Xiao L Feng J C Tian Y H Zhou S Q Zhou Y 2010 The Mechanisms of Resistance Spot Welding of Magnesium to Steel *Metallurgical and Materials Transactions A* **41** (10) 2651-2661
[14] Pouranvari M Alizadeh-Sh M and Marashi S 2015 Welding metallurgy of stainless steels during resistance spot welding Part I: fusion zone *Science and Technology of Welding and Joining* **206** 502-511
[15] Marashi P Pouranvari M Amirabdollahian S Abedi A and Goodarzi M 2008 Microstructure and failure behavior of dissimilar resistance spot welds between low carbon galvanized and austenitic stainless steels *Materials science and engineering: A* **4801** 175-180

IOP Conf. Series: Materials Science and Engineering **238** (2017) 012005 doi:10.1088/1757-899X/238/1/012005

The Effect of Dry and Wet Condition on the Mechanical Properties of Hybrid Single Lap

M Mariam, M Afendi and M S Abdul Majid

School of Mechatronic Engineering, Universiti Malaysia Perlis, Kampus Pauh Putra, Arau, Perlis.

E-mail: mariam.bt.majid@gmail.com

Abstract. The aim of this study is to investigate the strength and mechanical behaviour of similar and dissimilar materials AA7075/E-glass single lap joint. The specimen radially in stack with Araldite epoxy adhesive and mechanical fasteners of Huck bolt with the bondline length of 64 mm for different joint configurations; bolted, bonded and hybrid. The investigation was conducted under two different conditions of dry and wet of 55°C for long water immersion of 120 days. The effect of moisture was defined in hybrid joints as combination of adherend and bonding materials. The tensile test with a speed rate of 1 mm/min was conducted for both condition using 100 kN load cell until total failure of specimen. The experimental results were plotted into load displacement and stress strain curve for each environment. In addition, failure mechanism of the specimens was detected in which occurred at adhesive and Huck bolt.

1. Introduction

The joint designs are key to exploit maximum capacity of composite materials, which are as a rule progressively utilized as a part of aircraft structures. The bolted joints were preferred for joining the structures which suffer from low joint efficiencies. Ordinarily, a composite joint accomplishes a joint productivity of 40–50% contrasted with 60% for metals [1]. The time and cost effective maintenance had been resolved by introduce mechanical fasteners of screw, rivet, and bolts which are regularly utilized on the grounds. This allow taking into an action for non-destructive and simply reassembly of structures. The hybrid technology is a combination of adhesive and fasteners as bonding and bolting parameters which prove the best properties and performance for individual stiffness in each load path [2]. Andure M W reported that the fuel consumption in transportation parts can be reduced by achieving the high strength in components by hybrid joint [3]. Fu and Hoang-Ngoc demonstrated the load sharing using low modulus of adhesive and found that the bolt does not participate initially in sharing the load until the adhesive has completely yielded [4, 5].

The utilization of plastic materials in engineering structures has increased due to their low weight, high specific strength, high specific modulus, design flexibility, reduced manufacturing cost, excellent corrosion, solvent and environmental resistance, thermal and electrical insulation and durability [6]. Rather than that, environmental effect of heat and humidity at long period can be avoid as corrosion preventative measures are taken into fasteners for being insensitivity. The water may diffuse into metal adherend in joints by: (a) diffusion through the adhesive, (b) transport along the interface, and (c) capillary action through cracks and crazes in the adhesive.

In the present study, the effect of dry and wet condition on mechanical properties of single lap joint were investigated. Three different joint configuration: bolted, bonded, and hybrid were used to evaluate the mechanical properties under dry condition. Meanwhile, only hybrid single lap joint was

IOP Conf. Series: Materials Science and Engineering **238** (2017) 012005 doi:10.1088/1757-899X/238/1/012005

exposed in wet condition with a long period of water immersion of 120 days. The hybrid joints purposely exposing the adhesive layer to moisture and drilling holes at the center to accelerate the temperature. Thus, the experiment continues by comparing the mechanical behaviour for dry and wet environment through tensile test.

2. Materials and specimen preparations

2.1. Materials Selection

Aluminium alloy AA7075 and [0/90°] laminates of E-glass were used in this experiment with the dimension of (160 x 40 x 3) mm as shown in figure 1. The composites were laminates with 17 plies and 1:3 ratios of hardener and epoxy. The composite was examined in preliminary tensile test of E-glass dog bone shape design shown in figure 2 accordance with standard ASTM D3039. The extensometer with gauge length of 50 mm was attached on the specimen. The composite properties were examined to calculate and define the bonding area. The function of joining or overlap or bondline length in bolted, bonded and hybrid were defined the length by which the adherends will attached one another. These design parameter was applied as one of the most important factors affecting the joint rotation, stress distribution and ultimate tensile strength of single lap joint. The attachment area was calculated using formula determined by in Eq. 1 as shown. The mechanical properties of aluminium and glass fiber are tabulated in table 1. The bonding area was set to 64 mm. In part, table 2 show adhesive properties of Araldite at elevated temperature, contribute to 55°C in this experiment with elastic modulus of 0.426 GPa.

$$L = \sigma_y t / \tau \quad (1)$$

Where,

σ_y = Yield strength of GFRP composites

t = Thickness of the adhesive

τ = 50 % of the shear strength of adhesive

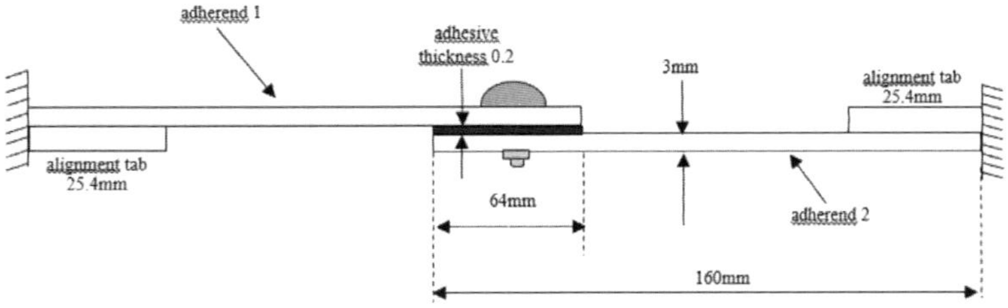

Figure 1. Hybrid single lap joint dimension.

IOP Conf. Series: Materials Science and Engineering **238** (2017) 012005 doi:10.1088/1757-899X/238/1/012005

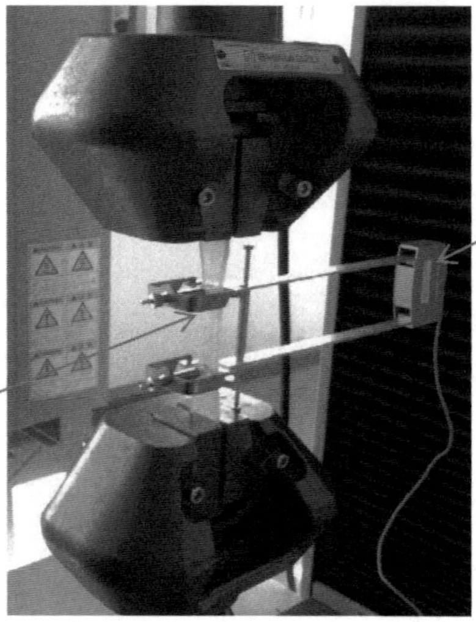

Figure 2. Dog bone shape preliminary tensile test.

Table 1. Mechanical properties of adherends.

Material	Tensile Strength, [MPa]	Poison ratio, v	Elongation, (%)
Aluminium alloy AA7075	**228**	**0.33**	**1.7**
E-glass	**157.052**	**0.19**	**2.3**

Table 2. Araldite epoxy adhesive properties.

Temperature, °C	Elastics Modulus, GPa	Poisson ratio, v
Room Temperature	1.392	0.380
55	0.426	0.463

2.2. Specimen preparations

The similar and dissimilar adherents of metallic alloy and composite were combined into bolted, bonded, and hybrid single lap joint for dry condition. Thus, mechanical joint strength obtained was compared with the wet condition on dissimilar hybrid single lap joint. According to figure 3, the flow of both condition was illustrated. In joining manufacturing, the jig was prepared and aluminium alloy was placed on the top as the bottom plate for the joint in first step. The jig was manufactured using black thermos material with the required shape for placing five specimens at one time. In the second step, the Araldite epoxy was applied on the top of aluminium along the bondline length of 64 mm and 0.2 mm thickness. The glass fiber plate was inserted on the top of adhesive. A small alignment tabs (25mm x 25mm) were inserted at the both end of adherends to assure a correct alignment and get symmetry during the test [7]. The specimens were then being rested for 24 hours. Properly, the Huck bolt types C6L Lockbolts was inserted using hydraulic installer at the centre of bond length with diameter hole of 6.4 mm in final step. The steps were repeated for bolted and bonded joint as shown in figure 4.

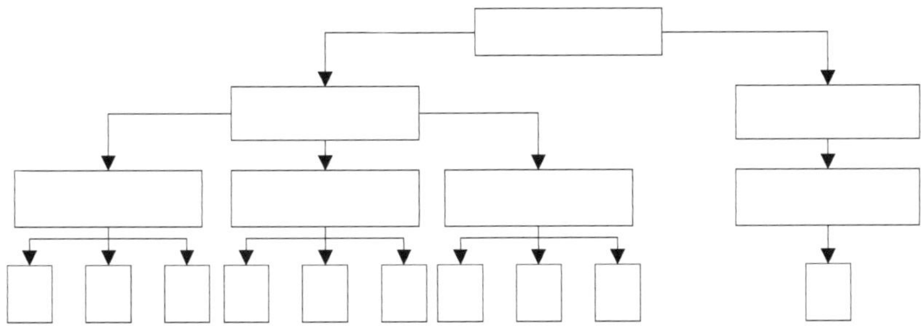

Figure 3. Flow chart of dry and wet condition: A (similar aluminium alloy), E (similar E-glass composite), AE (combination of aluminium and E-glass).

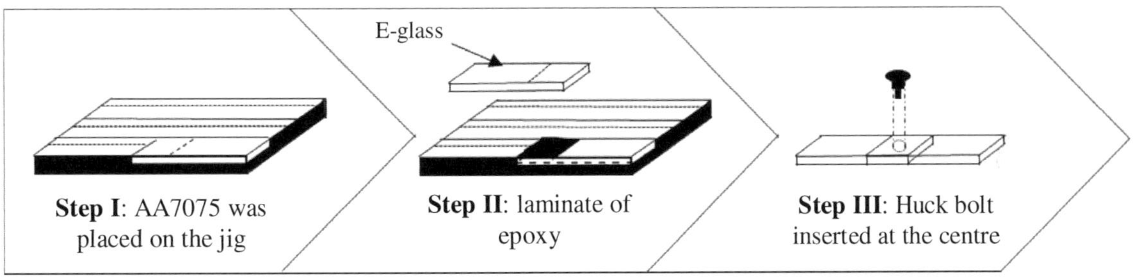

Figure 4. Manufacturing steps for hybrid joint.

3. Experimental Procedure

Figure 5 shows the 100 kN load cell used during tensile testing to determine the elastic-plastic mechanical properties of each joint configurations. Five specimens for each joining and conditions were experimentally conducted at speed rate set up to 1 mm/min with the gauge length of 206 mm. The composite specimen was placed on top as recommended for dissimilar materials joining testing. Furthermore, the head of Huck C6L lockbolt should be placed on the composite surface as lock tail on metal surface.

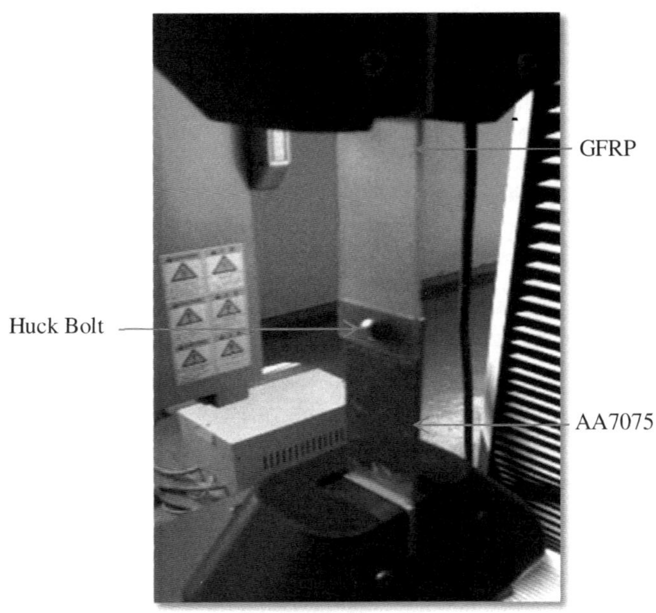

Figure 5. Tensile test

4. Results and Discussion

4.1. Loading Effect in Joint Configuration

The ultimate tensile failure load of bolted, bonded and hybrid joints were obtained from the test shown in figure 6. The average value of maximum load was plotted with similar and dissimilar materials for dry condition. It can be observed that, hybrid joining exhibits highest failure load followed by bonded and bolted joint for both similar and dissimilar materials. According to the error bars chart, dissimilar AA7075/E-glass demonstrates the highest strength with 18.34±0.61 kN, reducing the strength with similar E-glass and AA7075. The differences rate is 13.2% with increasing of joint strength. However, the strength in similar AA7075 raised up in bolted joint rather than hybrid. Two failure was detected in hybrid joint as adhesive failed at initial and mechanically fastened of Huck bolt support the joint and failed as final failure. The stress distribution is uniform in bonded joint, while the crack and fracture occurred when stress concentration arisen at the drilling hole. The maximum load for each joining was tabulated in table 3.

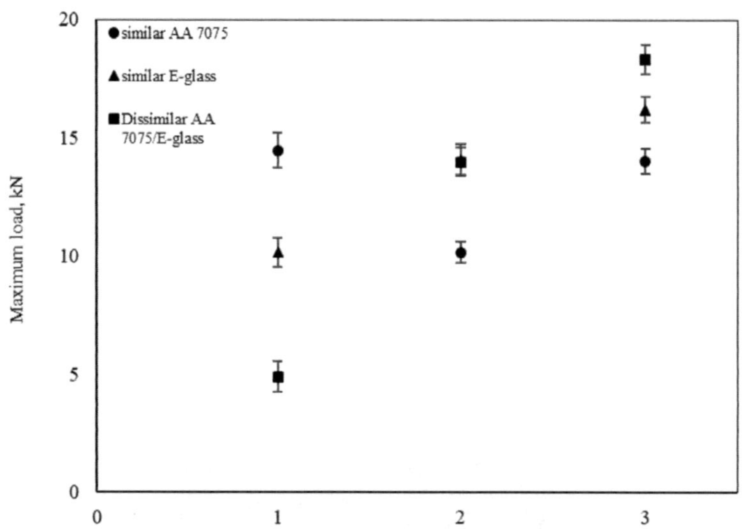

Figure 6. Maximum load in similar and dissimilar materials under dry condition for bolted (1), bonded (2) and hybrid (3) joints.

Table 3. Ultimate tensile failure load data.

Specimens	Ultimate Tensile Failure Load (kN)		
	Bolted	Bonded	Hybrid
Similar AA7075	14.47±0.73	10.16±0.44	14.02±0.51
Similar E-glass	10.14±0.63	14.08±0.65	16.20±0.53
Dissimilar AA7075/E-glass	4.90±0.65	13.99±0.611	18.34±0.61

4.2. Effect of Wet Environment On Hybrid Joint Strength

All the specimens prepared were conducted in tensile test to evaluate the strength performance with the load displacement and stress strain behaviour. The effect of dissimilar hybrid joint on wet condition was compared with dry condition as illustrated in figure 7. The plotting curved of the joints are based [0/90°] laminates of composites and Araldite adhesive. This clearly indicates that the tensile strength of long water immersion affected the joint behaviour and the reduction about 62% compared with the dry environment. The adhesive and Huck bolt failed at first and second fracture, respectively. Meanwhile, it can be observed that the adhesive failure occurred at first fillet and structural was failed after second fillet once the Huck bolt was significantly damaged. Mostly, the structural failure was observed on the composite specimen at the edge of bolt holes. The increasing of second fillet was attributed to load sharing between adhesive and Huck bolt. The ultimate strength of dry and wet is calculated by 17.729 kN and 6.819 kN, respectively. Meanwhile, the tensile strength of hybrid joint decreased from 63 MPa under dry conditions to 24 MPa under wet conditions; representing a decrease of over 60%.

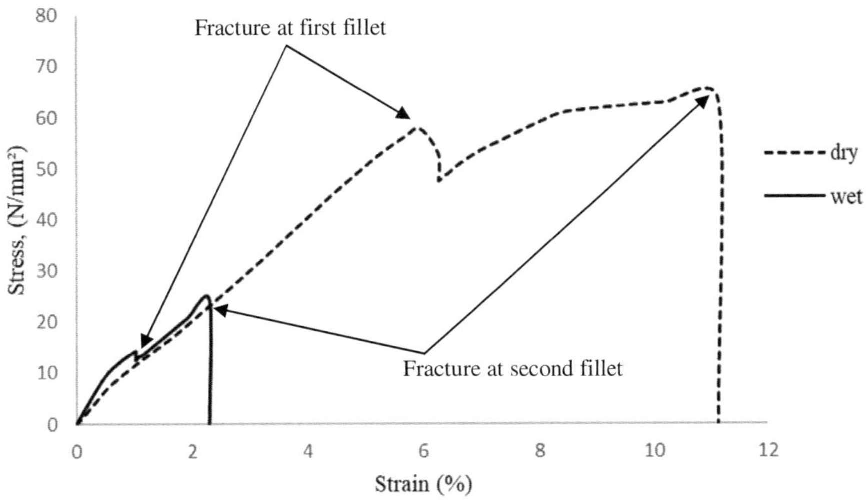

Figure 7. Stress strain behaviour.

4.3. Analysis of Failure Types

The adhesion and structural failure was identified in the study as in figure 8. The failed adhesive was stack on the lap side of both substrates. The failure was occurred by fiber tear in one of the substrates which can be described as adhesion failed. This was due to long moisture exposure with the water diffused in composite and adhesive. The bolt holes on composites were expanded before splitting and net tension until reach maximum loaded to fracture.

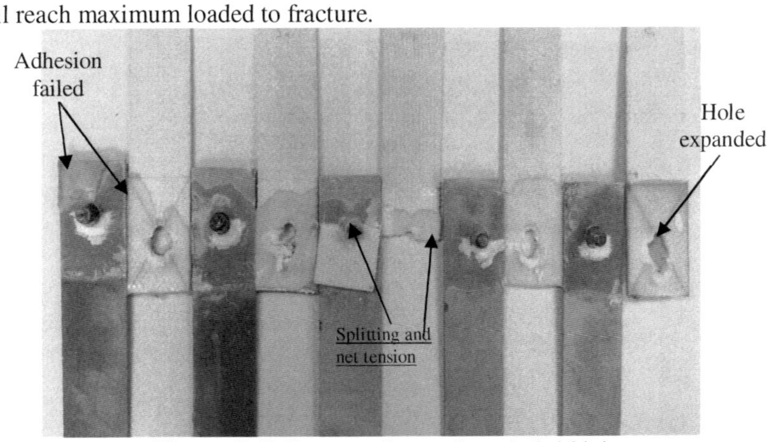

Figure 8. Failure specimen of wet hybrid joint.

5. Conclusion

The strength of hybrid joints under dry and wet environment had been investigated through experimental work. It can be concluded that, the water absorption at long exposure highly affects the strength of joint about 62% in wet condition. The highest failure mode for both conditions are 63 MPa and 24 MPa, for dry and wet, respectively. Also, the primary failure mode occurred in adhesive joint, and prevented by the presence of bolt as secondary failure before final separation.

References

[1] Raju K P Bodjona K Lim G H and & Lessard L 2016 Improving load sharing in hybrid bonded/bolted composite joints using an interference-fit bolt *Composite Structures* **149** 329-338

[2] Gómez S Oñoro J and Pecharromán J 2007 A simple mechanical model of a structural hybrid adhesive/riveted single lap joint *International Journal of Adhesion and Adhesives* **27** (4) 263-267

[3] Andure M W Jirapure S C D L P 2012 Advance Automobile Material for Light Weight Future – A Review *International Conference on Benchmarks in Engineering Science and Technology ICBEST 2012* Proceedings Published by International Journal of Computer Applications® (IJCA) 15-22

[4] Fu M and Mallick P K 2001 Fatigue of hybrid (adhesive/bolted) joints in SRIM composites. *International Journal of Adhesion and Adhesives* **21**(2) 145-149

[5] Hoang-Ngoc C T and Paroissien E 2010 Simulation of single-lap bonded and hybrid (bolted/bonded) joints with flexible adhesive *International Journal of Adhesion and Adhesives*, **30** (3) 117-129

[6] Samaei M Zehsaz M and Chakherlou T N 2016 Experimental and numerical study of fatigue crack growth of aluminum alloy 2024-T3 single lap simple bolted and hybrid (adhesive/bolted) joints *Engineering Failure Analysis* **59** 253-268

[7] Fiore V Alagna F Di Bella G and Valenza A 2013 On the mechanical behavior of BFRP to aluminum AA6086 mixed joints *Composites Part B: Engineering* **48** 79-87

Feasibility of using acoustic method in monitoring the penetration status during the Pulse Mode Laser Welding process

M F M Yusof [1,2], M Ishak[1], M F Ghazali[2]

[1]Joining & Welding Research Group (J&W), Faculty of Mechanical Engineering, Universiti Malaysia Pahang (UMP),26600 Pekan, Pahang, Malaysia,

[2]Advanced Structural Integrity & Vibration Research Group (ASIVR), Faculty of Mechanical Engineering, Universiti Malaysia Pahang (UMP),26600 Pekan, Pahang, Malaysia,

E-mail: fadhlan@ump.edu.my

Abstract. In this paper, the feasibility of using acoustic method to monitor the depth of penetration was investigated by determine the characteristic of the acquired sound throughout the pulse mode laser welding process. To achieve the aim, the sound signal was acquired during the pulsed laser welding process on the 2 mm structural carbon steel plate. During the experiment, the laser peak power and pulse width was set to be varied while welding speed was constantly at 2 mm/s. Result from the experiment revealed that the sound pressure level of the acquired sound was linearly related to the pulse energy as well as the depth of penetration for welding process using 2ms pulse width. However, as the pulse width increase, the sound pressure level show insignificant change with respect to the change in the depth of penetration when the pulse energy reaches certain values. The reported result shows that this was happen due to the occurrence of spatter which suppressed the information associated with the generation of plasma plume as the product of high pulse energy. In this work, it was demonstrated that in some condition, the acoustic method was found to be potentially suitable to be used as a medium to monitor the depth of weld on online basis. To increase the robustness of this method to be used in wider range of parameter, it was believed that some other post processing method is needed in order to extract the specific information associated with the depth of penetration from the acquired sound.

1. Introduction

In recent years, there have been increasing demands on the laser welding process in wide variety of industries due to its ability in promoting small heat affected zone, aesthetic appearance, less post-weld machining process and high production rate [1]. Accordingly, numerous studies associated with laser welding process have been recorded since past several decades. One of the problem angle emphasized by many researchers are the optimum parameters for the process. However, many scholars hold the view that the studies on the robust monitoring method for the process are also essential. According to [2], process monitoring is important because it could promote greater control during the process. Over the past several years, many methods inclusive of electrical, thermal, optical as well as acoustic method for monitoring laser welding have been studied. Among those methods, acoustic method received some great attention recently due to low cost, simple, high responsible speed and convenient features [3-5].

In many industries, defect such as incomplete penetration is intolerable because it might affect the strength of the component. Therefore, over a few decades, most of the researcher had tried to find the deeper understanding on the feasibility of using the acoustic method to monitor the penetration status during the laser welding process. In early stage, the correlation between acoustic spectral features and depth of penetration during the laser welding process on Aluminum 1100 have been investigated by Duly and Mao [6]. According to the result, it was reported that peak within 9 kHz to 10 kHz was increasing in its amplitude with gaining laser intensity which simultaneously increased the depth of penetration. Similar work also have been reported by Farson et al., [7] whereas the experiment was done on 304 stainless steel plate. However, the significant range was recorded to be within 1 kHz to 2 kHz in which its energy drops when insufficient penetration detected. In the extended study by Farson et al., [8], the investigation was done in broader range of power and travel speed. It was describe that the Root Mean Square (RMS) of acoustic signal was high at moderate full penetration and it was decreasing as the penetration status felt into partial penetration class. In other research involving high power or keyhole welding, Farson et al., [9] reported that the depth of penetration status could be classified into full penetration, overheat penetration and half penetration. In this study, the overheat penetration was referred to weld produced by excessive linear heat input which consequently caused a significant level of top surface concavity as well as larger heat affected zone. Unlike the other studies, Huang and Kovacevic [3], have demonstrated the use of spectral subtraction method prior to classification process in order to diminish the influence of noise during laser welding. It was described that the overall sound pressure increases align with the growing in penetration depth whilst the power density of frequency spectrum at 500 Hz to 1500 Hz was large for full penetration case as compared to half penetration. Since classifying the penetration status is insufficient for the case where the depth of penetration needs to be quantitatively measured, Huang & Kovacevic [10] have utilized the Artificial Neural Network analysis in attempt to quantitatively characterize the relation between the captured sound and the depth of penetration.

Commonly, in various laser welding application, the process was done either in continuous or pulse mode. Between both wave mode type, pulse wave mode seems to be advantageous due to several reason. By using pulse wave mode, high average power laser machine are unnecessary [1] in order to achieve high penetration. Moreover, pulse laser welding promotes very low heat input to the weld, which gives low distortion as well as giving more chances to weld heat-sensitive components [11-15]. Up to now, far too little attention has been paid to the study of sound behavior during the pulse laser welding. It is believed that the understanding on the sound behavior emitted during the pulse laser process would widen the option in using acoustic method for monitoring the weld penetration on real-time basis. In this paper, the characteristic of the acquired sound throughout the pulse mode laser welding process will be determine in order to investigate the feasibility of using this method to monitor the depth of penetration. The relation between sound pressure levels of the acquired sound with the depth of penetration will be presented and further discussed.

2. Experimental Procedure

In order to establish the relation between sound emitted from the pulse laser welding process and the depth of penetration, the experiment was set up as illustrated in figure 1. Basically, the bead on plate laser welding process was done onto the structural carbon steel specimen with dimension of 25 x 60 x 2 mm. Prior to the experimental work, each of the specimen was sanded using 120 grid sand paper intently to increase the absorptivity of the laser energy. The peak power and pulse width were set to be varied according to table 1. The focal length and angle used in this experiment were fixed to 148 mm and 50 respectively. For each of the parameter set, the experiment was repeated 10 times. During the welding process, the specimen move with constant speed of 2 mm/s while the argon gas with flowrate of 15 L/min was used as shielding gas.

Meanwhile, the emitted sound was acquired using PCB piezotronics free-field microphone which capable of capturing the flat response signal between 20 Hz to 10 000 Hz. Prior to the acquisition process, the sampling rate was set to be 25.6 k. Sample/s which was determine according to Nyquist theorem. Meanwhile, the sound pressure level, SPL was determine using formula as stated in equation (1) whereas P and Pref are the sound pressure and pressure reference respectively.

$$SPL = 20 \log_{10} (P / P_{ref}) \quad (1)$$

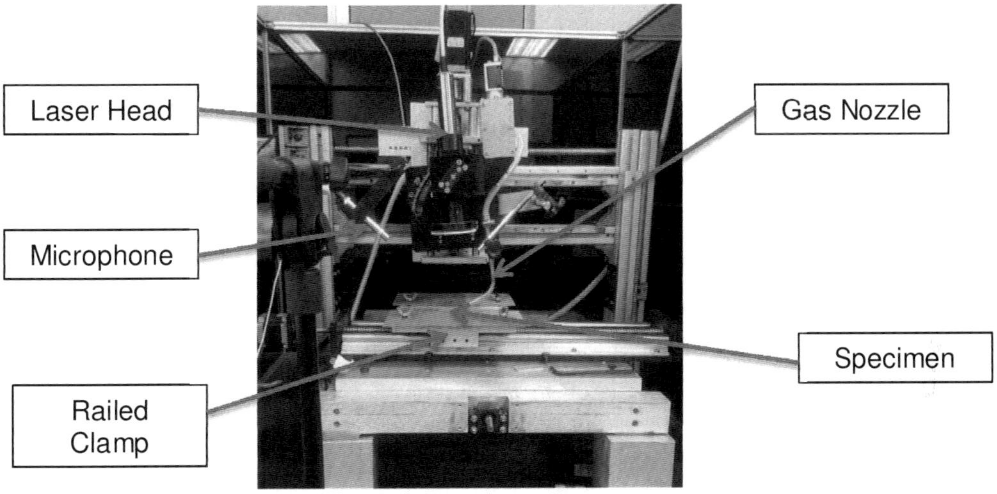

Figure 1. Experiment Setup.

Table 1. Parameter variation for pulse laser welding process.

Pulse width (ms)	Laser Peak Power (W)	Pulse width (ms)	Laser Peak Power (W)	Pulse width (ms)	Laser Peak Power (W)
	400		400		400
	600		600		600
	800		800		800
2	1000	4	1000	6	1000
	1200		1200		1200
	1400		1400		1400
	1600		1600		1600

3. Results and Discussions

In this work, the first question aimed to look into the acoustic signal behavior originated from laser welding process. As illustrated in figure 2, for laser welding with pulse width of 2 ms, it is apparent that the overall amplitude increased with gaining value of peak power. This trend further support the idea from the previous work [16, 17] whereas it was reported to happen because of the plasma plume size produced from the laser welding process with different peak power. According to Hoffman et al [2], the pressure produced from the plasma plume could produce sound. High air pressure produce sound with large overall amplitude and the amount of air pressure is depending on the plume size whilst it is rely upon the peak power.

As comparison was made with the result from the welding process with pulse width of 4 ms and 6 ms in figure 3 and 4 respectively, the striking observation emerged from the overall amplitude value. For both cases, the overall amplitude of sound barely changes at certain value of peak power. No significant change was recorded on the overall amplitude of sound signal when peak power was

IOP Conf. Series: Materials Science and Engineering **238** (2017) 012006 doi:10.1088/1757-899X/238/1/012006

1200W and above for welding process with pulse width of 4ms. Similar pattern also occur for laser welding process with pulse width of 6 ms when peak power was recorded to be above 800W. A possible explanation for the result trend might be found from the image captured during the welding process as presented in figure 5 to 7. Referring to the figure 5, with pulse width of 2ms, high amount of spatter occurs only when the peak power reach the maximum value. As the pulse width increase, spatter starts to take place earlier, For instance, based on the captured image, spatter was detected when the peak power reach 1200W and 600W for the case of 4ms and 6ms pulse width respectively. This could be discovered from figure 6 and 7. According to Dawes [1], one of the reasons of the spatter occurrence is the excessive laser power. Hence, by looking into the trend of spatter occurrence from this this experiment, it is believed that barely changed overall amplitude of the acquired sound was probably due to the large amount of noise produced by it. In essence, spatter could occur due to instability of keyhole caused by large amount of pressure from the evaporation of melt surface [18]. Meanwhile, in case of pulse laser welding, the amount of melt surface evaporation is relying on the laser energy which influenced by both peak power and pulse width together.

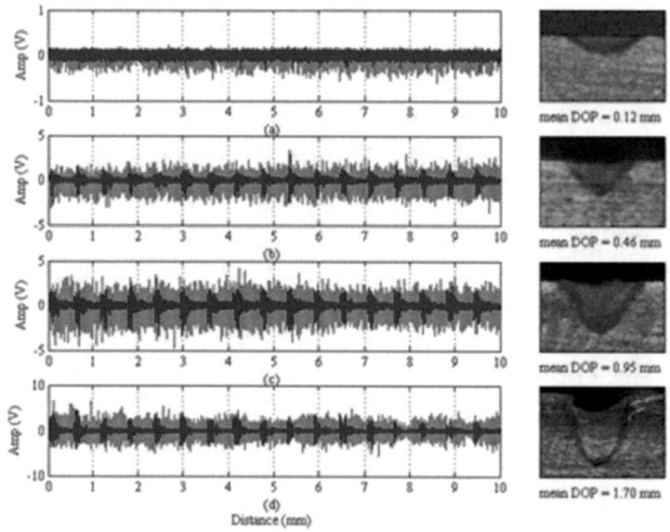

Figure 2. Sound signal acquired from welding process with pulse width of 2 ms and peak power (a) 400W (b) 800W (c) 1200W (d) 1600W.

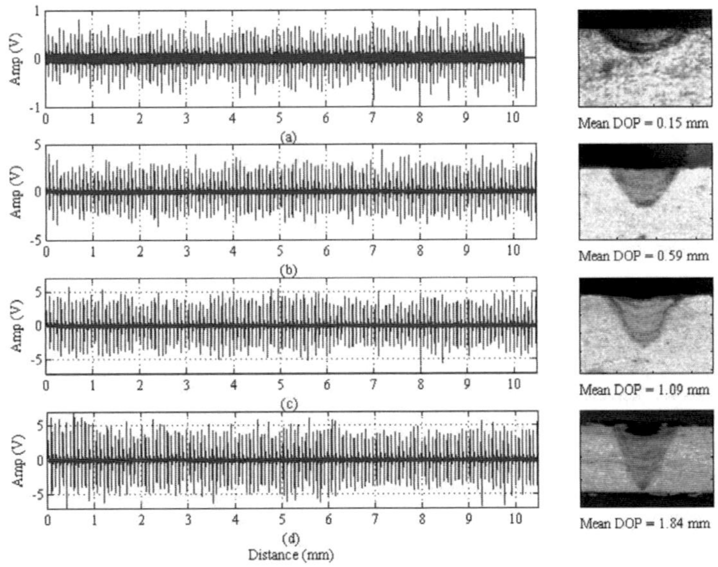

Figure 3. Sound signal acquired from welding process with pulse width of 4 ms and peak power (a) 400W (b) 800W (c) 1200W (d) 1600W.

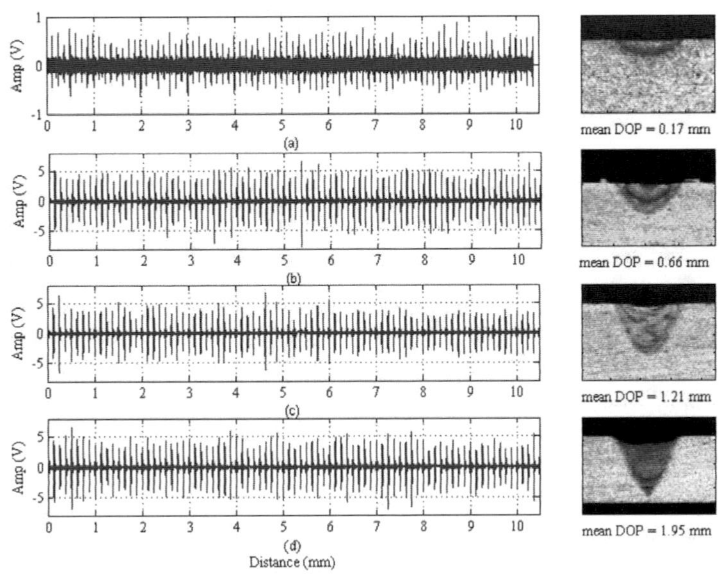

Figure 4. Sound signal acquired from welding process with pulse width of 6 ms and peak power (a) 400W (b) 800W (c) 1200W (d) 1600W.

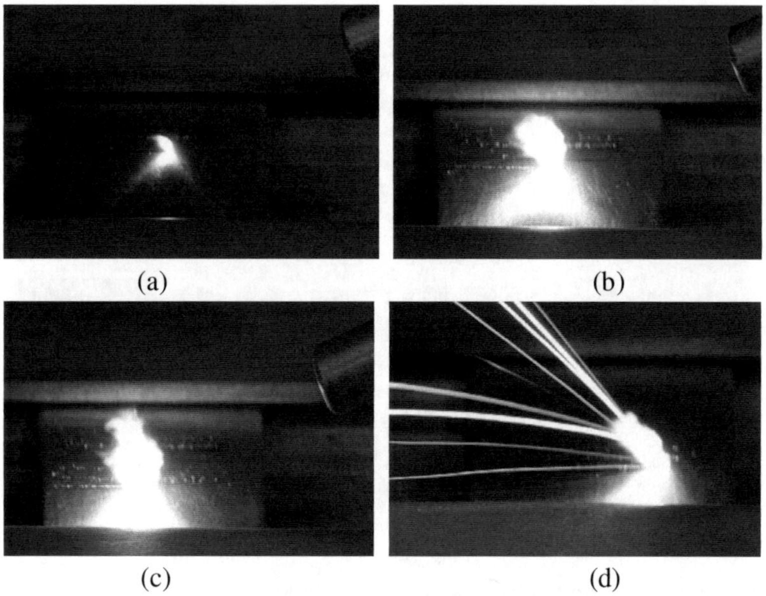

Figure 5. Laser welding process with pulse width of 2ms and peak power (a) 400W (b) 800W (c) 1200W (d) 1600W.

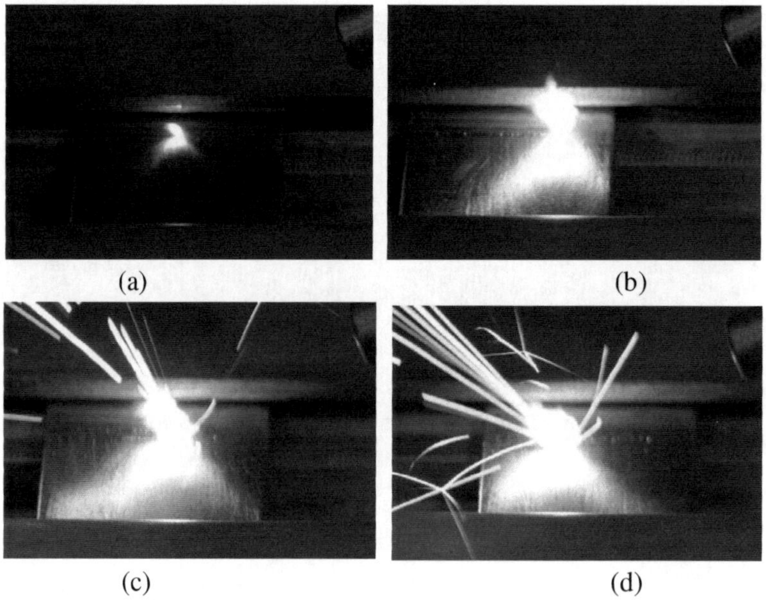

Figure 6. Laser welding process with pulse width of 4ms and peak power (a) 400W (b) 800W (c) 1200W (d) 1600W.

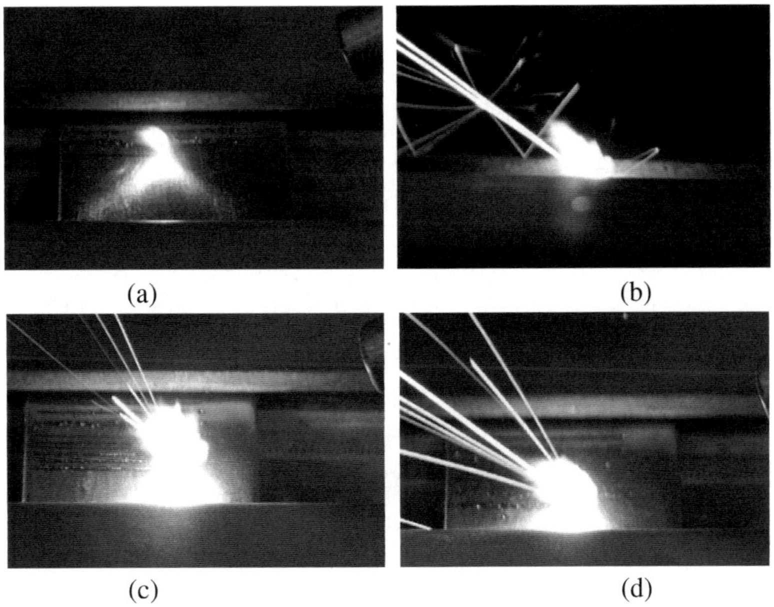

Figure 7. Laser welding process with pulse width of 6ms and peak power (a) 400W (b) 800W (c) 1200W (d) 1600W.

To show the overall relation between the acquired sound pressure, depth of penetration, as well a the laser energy, the plot between laser energy and depth of penetration was plotted as shown in figure 8 while the colerration between laser energy and sound pressure level was illustrated in figure 9. Both relations was plotted separatedly according to pulse width due to the reason that there will be a significant different in weld geometry cause by changing in pulse shape. Meanwhile, plotted sound pressure level was the average value from 10 repetitions for each experiment. As could be observed from the figure 8, conclusion could be drawn from the overall trend of the relation between laser pulse energy and depth of weld. It was evident that the depth of penetration is directly related with the laser pulse energy. However, contradict trend was recorded when relation between laser pulse energy and sound pressure level was made in figure 9. It could be obviously noticed that the sound pressure level values were slightly inconsistent when pulse energy reach certain values. This could be seen when pulse energy was 4 J and 4.8 J for the case of welding with pulse width of 4ms and 6ms respectively. Moreover, for both cases, the fluctuation range was within 1 dB to 2 dB. If the information from figure 8 and figure 9 was combined, it could be acknowledge that at pulse width and pulse energy was 4 ms and 4J, the depth of penetration just reach 0.99 mm. Meanwhile, the depth of penetration was recorded to be 0.66mm when pulse energy was recorded to be 4.8 J for the case of laser welding process with pulse width of 6 ms. As reported in the previous study [16, 17] the sound pressure would increase with the larger value of pulse energy. This is due to the amount of plasma plume that influenced by the pulse energy. However, in this study, the occurrence of spatter might suppress the sound generated by the plasma plume. Consequently, the trend of sound pressure level with increasing amount of pulse energy as well as the penetration depth was unclear at the moment the spatter starts to occur. In other words, insignificant trend was detected when the process are unstable.

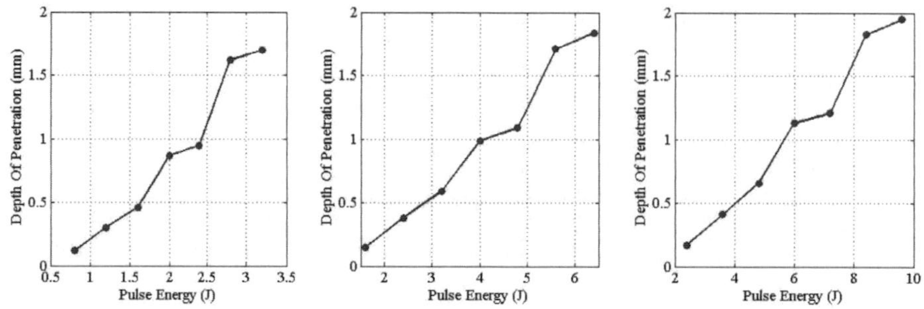

(a) (b) (c)

Figure 8. Relation between laser pulse energy and weld depth
(a) Pulse width 2ms (b) Pulse width 4ms (c) Pulse width 6ms.

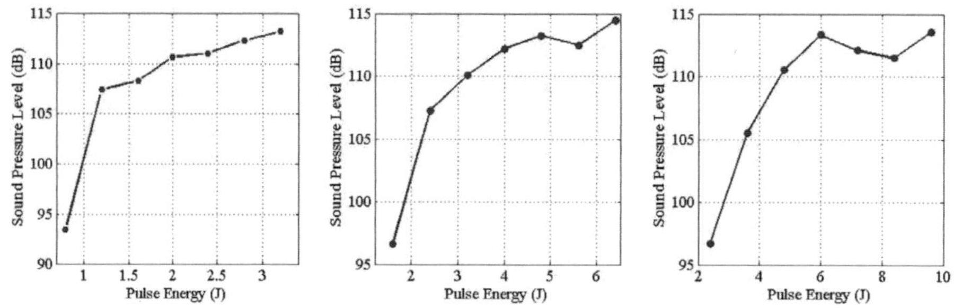

(a) (b) (c)

Figure 9. Relation between laser pulse energy and sound pressure level
(a) Pulse width 2ms (b) Pulse width 4ms (c) Pulse width 6ms .

4. Conclusion

In this work, the feasibility of using acoustic signal to monitor the depth of penetration for the case of pulse laser welding was investigated. Result from the experiment revealed that the sound pressure level of the acquired sound was linearly related to the pulse energy as well as the depth of penetration for welding process using 2ms pulse width. However, as the pulse width increases, the sound pressure level show insignificant change with respect to the change in pulse energy as well as the depth of penetration when the pulse energy reaches certain values. The reported result shows that this was happen due to the occurrence of spatter which suppressed the information associated with the generation of plasma plume as the product of high pulse energy.

To conclude, the acoustic method was found to be potentially suitable to be used as a medium to monitor the depth of weld on online basis. However, some other post processing is needed in order to eliminate the influence of noise which affecting the trend of the result.

References

[1] Dawes C 2008 Laser Welding A Practical Guide. United Kingdom (Abington Publishing)

[2] Hoffman J, Szymanski Z, Jakubowski J and Kolasa A 2002 Analysis of acoustic and optical signals used as a basis for controlling laser welding processes *Welding International* 2002/01/01. **16** 18-25

[3] Ao S, Luo Z, Feng M and Yan F 2015 Simulation and experimental analysis of acoustic signal characteristics in laser welding *The International Journal of Advanced Manufacturing Technology* **81** 277-287

[4] Huang W and Kovacevic R 2009 Feasibility study of using acoustic signals for online monitoring of the depth of weld in the laser welding of high-strength steels *Proceedings of the Institution of Mechanical Engineers, Part B: Journal of Engineering Manufacture* **223** 343-361

[5] Luo Z, Liu W, Wang Z and Ao S 2016 Monitoring of laser welding using source localization and tracking processing by microphone array *The International Journal of Advanced Manufacturing Technology* **86** 21-28
[6] Duley W W and Mao Y L 1994 The effect of surface condition on acoustic emission during welding of aluminium with CO 2 laser radiation *Journal of Physics D: Applied Physics* **27** 1379
[7] Farson D, Hillsley K, Sames J and Young R 1996 Frequency–time characteristics of airborne signals from laser welds *Journal of Laser Applications* **8** 33-42
[8] Farson D, Ali A and Sang Y 1998 Relationship of optical and acoustic emissions to laser weld penetration *Welding journal* **77** 142. s-148 s
[9] Farson D, Ali A and Li X 1999 Laser weld penetration monitoring with multiple emission signal measurements *Journal of Laser Applications* **11** 47-53
[10] Huang W and Kovacevic R 2011 A neural network and multiple regression method for the characterization of the depth of weld penetration in laser welding based on acoustic signatures *Journal of Intelligent Manufacturing* **22** 131-143
[11] Frewin M and Scott D 1999 Finite element model of pulsed laser welding *WELDING JOURNAL-NEW YORK-* **78** 15-s
[12] Wu D, Ma G, Niu F and Guo D 2013 Pulsed laser welding of Hastelloy C-276: high-temperature mechanical properties and microstructure *Materials and Manufacturing Processes* **28** 524-528
[13] Assuncao E and Williams S 2013 Comparison of continuous wave and pulsed wave laser welding effects *Optics and Lasers in Engineering* **51** 674-680
[14] Liu J, Weckman D and Kerr H 1993 The effects of process variables on pulsed Nd: YAG laser spot welds: Part I. AISI 409 stainless steel *Metallurgical and Materials Transactions B* **24** 1065-1076
[15] Weckman D, Kerr H and Liu J 1997 The effects of process variables on pulsed Nd: YAG laser spot welds: Part II. AA 1100 aluminum and comparison to AISI 409 stainless steel *Metallurgical and Materials Transactions B* **28** 687-700
[16] Conesa S, Palanco S and Laserna J J 2004 Acoustic and optical emission during laser-induced plasma formation *Spectrochimica Acta Part B: Atomic Spectroscopy* **59** 1395-1401
[17] Khosroshahi M E, Pour F A, Hadavi M and Mahmoodi M 2010 In situ monitoring the pulse CO2 laser interaction with 316-L stainless steel using acoustical signals and plasma analysis *Applied Surface Science* **256** 7421-7427
[18] Kaplan A and Powell J 2011 Spatter in laser welding *Journal of Laser Applications* **23** 032005.

IOP Conf. Series: Materials Science and Engineering **238** (2017) 012007 doi:10.1088/1757-899X/238/1/012007

Hardness variation of welded boron steel using continuous wave (CW) and pulse wave (PW) mode of fiber laser

K I Yaakob, M Ishak, S R A Idris M H Aiman and N Z Khalil

Faculty of Mechanical Engineering, Universiti Malaysia Pahang, 26600 Pekan, Pahang, Malaysia

E-mail: mahadzir@ump.edu.my

Abstract. Recent car manufacturer requirement in lightweight and optimum safety lead to utilization of boron steel with tailor welded blank approach. Laser welding process in tailor welded blank (TWB) production can be applied in continuous wave (CW) of pulse wave (PW) which produce different thermal experience in welded area. Instead of microstructure identification, hardness properties also can determine the behavior of weld area. In this paper, hardness variation of welded boron steel using PW and CW mode is investigated. Welding process is conducted using similar average power for both welding mode. Hardness variation across weld area is observed. The result shows similar hardness pattern across weld area for both welding mode. Hardness degradation at fusion zone (FZ) is due to ferrite formation existence from high heat input applied. With additional slower cooling rate for CW mode, the hardness degradation is become obvious. The normal variation of hardness behavior with PW mode might lead to good strength.

1. Introduction

In fulfilling the car manufacturer requirement for producing lightweight chassis with high safety features, boron steel is introduced replacing the conventional high strength steel (HSS). Apart from exhibiting high strength properties after hot press forming process, the springback issue also have been solved, allowing complex geometry parts design [1–10]. Poor energy absorbtion of boron steel promotes the tailor welded blank (TWB) technology in optimization of weight while maintaining the crashworthiness of the component [7,11–17]. Due to involvement of laser welding method in TWB, the comparative study between different types of laser welding conducted by Assuncau revealed that the fiber laser results higher efficiency and productivity. It was also found that the pulse wave (PW) mode is prominent in higher penetration efficiency compared to continuous (CW) under similar condition.

PW mode is where the laser produces short pulses and less amount of heat is transferred to the work piece. The advantages of PW when compared to CW mode includes the lower heat input, lower distortion, and higher efficiency [18–24]. The continuous weld also can be made by overlapping every spot of pulses from PW mode to form a fusion zone or seam is known as pulse seam welding. Contrast with CW mode which laser applied on the surface, create the molten pool and drag along the weld line according to welding speed. The welding speed of CW mode is inversely proportional to the width of fusion zone (FZ) and heat affected zone (HAZ) [25].

Both welding mode applied power density at various points in the weld zone may experience different thermal events along the weld area. It leads to various microstructure formation in the FZ and HAZ. The different microstructure lead to different mechanical properties and required further

IOP Conf. Series: Materials Science and Engineering **238** (2017) 012007 doi:10.1088/1757-899X/238/1/012007

investigation. However, most microstructures form in a small area and thus rather difficult to be observed. By determined the hardness properties of microstructure the difference of microstructure behavior can be determined instead of microstructure observation. It is also expected that the variation of hardness across the weld area also can be used to predict the weld area separation depending on its changes. Therefore, the aim for this work is to determine the hardness variation of welded boron steel. The comparison of hardness variation between CW and PW mode was made and discussed.

2. Experimental Procedure

This study was conducted by using 1.6 mm thick of uncoated 22MnB5 boron steel. The specimens were cut to 50 mm x 50 mm dimension sufficient for bead on plate (BOP) welding. The IPG-YLM-QCW series of a fiber laser with an average power of 200 W was used in CW and PW mode. Average power and welding speed for were fixed as 100 W of average power and 5 mm/s respectively for both welding mode. Additional 1 ms of pulse width and 60 Hz of pulse repetition rate were fixed for PW welding. The focus point was set as 15 mm below the specimen surface together with 15° of incident angle to avoid beam reflection. 15 L/min of Argon gas flow rate was use to protect the weld area blown by copper tube nozzle. The diameter the nozzle was 6 mm and the nozzle tip distance to beam were 30 mm with 20° of angle.

Table 1. Chemical compositions of boron steel (22MnB5).

Composition (%)	C	Si	Mn	P	S	Cr	Ni	B
Boron Steel	0.26	0.30	1.24	0.016	0.003	0.20	0.016	0.004

Figure 1. Welding process apparatus using fibre laser machine.

For the welding cross section observation, the specimens were finely cut and mounted. The mounted specimens were grounded, polished and etched using 2% of nital solution to reveal the microstructure difference. The observation of welded cross section area were carried out by the mean of optical microscope. The hardness test was conducted in a line across the weld area. The difference of hardness behavior across welded area between CW and PW mode were discussed.

3. Results and Discussion

Figure 2 shows the BOP cross section of CW and PW mode. It obviously shows that PW produce deeper penetration compared to CW mode. The penetration difference was due to the peak power density applied on the specimen by different welding mode. Even though both welding mode utilized similar average power, PW mode able to produce peak power. Meanwhile, CW mode only utilized average power as power output. Thus, the peak power density of PW mode is higher compared to CW mode which create deeper penetration. In another point of view, interaction time is the heating time of the process on weld centerline which also might affect the penetration. However, high interaction time of CW mode only affect the HAZ width compared to penetration. Thus, the HAZ of CW mode is wider compared to PW mode which produce shorter interaction time.

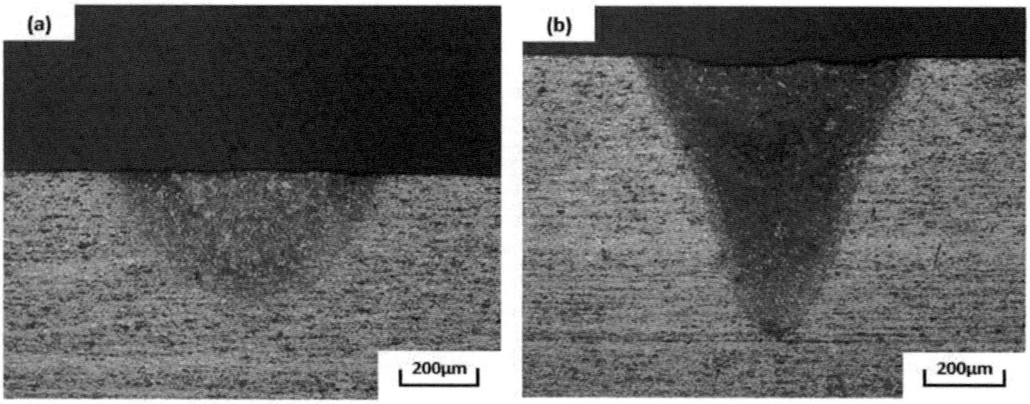

Figure 2. Welding cross section of (a) continuous wave and (b) pulse wave.

Figure 3 shows the hardness pattern across the weld area of CW and PW mode. CW mode produce wider weld area compared to PW mode. Both welding mode shows reduction of hardness value at the center of the weld. However, CW mode resulted wider weld area and huge hardness reduction compared to PW mode. This hardness variation might due to the microstructure transformation according to heat input and cooling rate during welding process. During welding process, the heat applied at FZ area high enough to melt the base metal. Followed by decrease in temperature during solidification might transform the base metal microstructure to martensite. However, the high temperature applied will enhance the boron element to rally at the boundary between FZ and HAZ. The boron concentration at this area might resulted borocarbide precipitation which promoting ferrite nucleation. Thus, the hardness of this area might reduce.

The heat from FZ will spread to the base metal and create HAZ area. HAZ consist of various microstructure due to temperature gradient moving from FZ to base metal. The temperature applied at HAZ might below the melting temperature. However, HAZ area near to FZ might experience the austenization temperature, Ac3 which optimum temperature for boron steel properties transformation. According to Naderi [26], starting from austenite temperature, 25 °C/s is the optimum cooling rate to obtain full transformation of boron steel properties to martensite microstructure. Thus, there are some HAZ area might result full martensite transformation.

The hardness of CW mode welding at FZ is lower than PW mode. During CW mode welding, FZ experience slower cooling rate compared to PW mode. It is because of the molten pool of CW mode is dragged along weld area which delay the solidification process and also reduce its cooling rate. The slower cooling rate also enhance the chances of borocarbide precipitation and ferrite nucleation which lead to hardness degradation. Contrast with PW mode welding which solidified faster in each pulse. Thus, the hardness degradation is not obvious compared to CW mode. HAZ hardness for both welding mode in the range of 500 to 600 HV. Even though the HAZ hardness of CW mode is slightly higher compared to PW mode, both welding mode might produce optimum temperature and cooling rate to produce full martensite transformation. However, HAZ of CW mode is wider than PW mode. It is due to slow cooling rate from FZ give sufficient time for heat to spread and create wider HAZ.

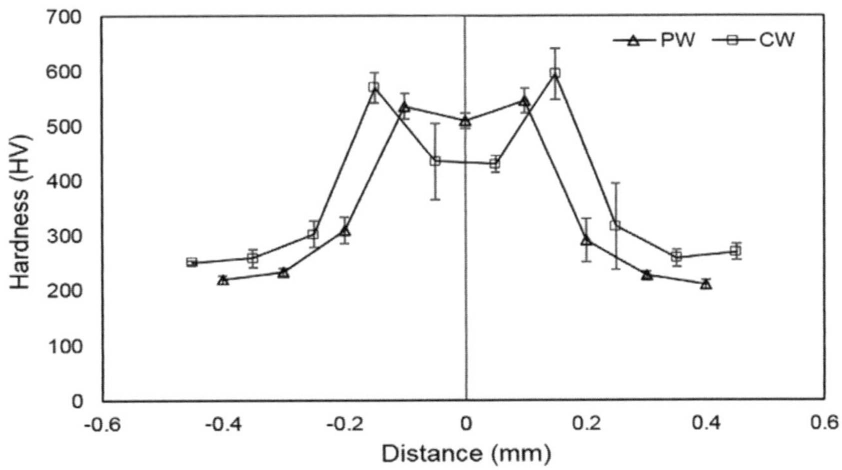

Figure 3. Hardness variation across weld area between continuous wave and pulse wave mode.

4. Conclusion

In this study, the investigation of hardness variation of welded boron steel on CW and PW mode have been conducted. The conclusion can be summarized as follows:

i. PW mode has better hardness distribution across weld area might lead to good strength.

ii. The hardness width of CW mode is determined by high interaction time sufficient for wider temperature movement. The fast interaction time and also rapid cooling from PW mode create smaller HAZ width.

iii. High hardness at HAZ for both welding mode is due to optimum thermal experience transform boron steel to AHSS properties. Low hardness at FZ is due to high temperature applied for melt the surface lead to ferrite formation. Obvious hardness degradation observed for CW mode due to slow cooling rate related to molten pool movement.

References

[1] Múnera D D, Pic A, Abou-khalil D and Shmit F 2008 Innovative Press Hardened Steel Based Laser Welded Blanks Solutions for Weight Savings and Crash Safety Improvements SAE *Int. J. Mater. Manuf.* **1** 472–9

[2] Maggi S and Murgia M 2008 Introduction to the metallurgic characteristics of advanced high-strength steels for automobile applications *Weld. Int.* **22** 610–8

[3] Naderi M, Ketabchi M, Abbasi M and Bleck W 2011 Analysis of microstructure and mechanical properties of different boron and non-boron alloyed steels after being hot stamped *Procedia Eng.* **10** 460–5

[4] Tang B T, Bruschi S, Ghiotti A. and Bariani P F 2014 Numerical modelling of the tailored tempering process applied to 22MnB5 sheets *Finite Elem. Anal. Des.* **81** 69–81

[5] Caron E J F R, Daun K J and Wells M a. 2014 Experimental heat transfer coefficient measurements during hot forming die quenching of boron steel at high temperatures *Int. J. Heat Mass Transf.* **71** 396–404

[6] Naderi M, Abbasi M and Saeed-Akbari a. 2012 Enhanced Mechanical Properties of a Hot-Stamped Advanced High-Strength Steel via Tempering Treatment *Metall. Mater. Trans. A* **44** 1852–61

[7] Merklein M, Wieland M, Lechner M, Bruschi S and Ghiotti A 2016 Hot stamping of boron steel sheets with tailored properties: A review *J. Mater. Process. Technol.* **228** 11–24

[8] Karbasian H and Tekkaya A E 2010 A review on hot stamping *J. Mater. Process. Technol.* **210** 2103–18

[9] Mori K I 2012 Smart hot stamping of ultra-high strength steel parts *Trans. Nonferrous Met. Soc. China (English Ed.* **22** 496–503
[10] Xing Z W, Bao J and Yang Y Y 2009 Numerical simulation of hot stamping of quenchable boron steel *Mater. Sci. Eng. A* **499** 28–31
[11] Tang B, Yuan Z, Cheng G, Huang L, Zheng W and Xie H 2013 Experimental verification of tailor welded joining partners for hot stamping and analytical modeling of TWBs rheological constitutive in austenitic state *Mater. Sci. Eng. A* **585** 304–18
[12] Xu F and Wang C 2016 Dynamic axial crashing of tailor-welded blanks (TWBs) thin-walled structures with top-hat shaped section *Adv. Eng. Softw.* **96** 70–82
[13] Merklein M, Johannes M, Lechner M and Kuppert A 2014 A review on tailored blanks—Production, applications and evaluation *J. Mater. Process. Technol.* **214** 151–64
[14] Xu F, Sun G, Li G and Li Q 2014 Experimental investigation on high strength steel (HSS) tailor-welded blanks (TWBs) *J. Mater. Process. Technol.* **214** 925–35
[15] Kang M, Kim Y-M and Kim C 2016 Effect of heating parameters on laser welded tailored blanks of hot press forming steel *J. Mater. Process. Technology* **228** 137-144
[16] Chandra D, Biro E, Gerlich A P and Zhou N Y 2016 Fusion zone microstructure evolution of fi ber laser welded press-hardened steels *Scr. Mater.* **121** 18–22
[17] Gerhards B, Reisgen U and Olschok S 2016 Laser Welding of Ultrahigh Strength Steels at Subzero Temperatures *Phys. Procedia* **83** 352–61
[18] Tzeng Y-F 2000 Process Characterisation of Pulsed Nd:YAG Laser Seam Welding *Int. J. Adv. Manuf. Technol.* **16** 10–8
[19] Sabbaghzadeh J, Azizi M and Torkamany M J 2008 Numerical and experimental investigation of seam welding with a pulsed laser *Opt. Laser Technol.* **40** 289–96
[20] Chelladurai A M, Gopal K a., Murugan S, Venugopal S and Jayakumar T 2014 Energy Transfer Modes in Pulsed Laser Seam Welding *Mater. Manuf. Process.* **30** 162–8
[21] Tadamalle A P, Reddy Y P and Ramjee E 2013 Influence of laser welding process parameters on weld pool geometry and duty cycle *Adv. Prod. Eng. Manag.* **8** 52–60
[22] Torkamany M J, Hamedi M J, Malek F and Sabbaghzadeh J 2006 The effect of process parameters on keyhole welding with a 400 W Nd : YAG pulsed laser *J. Phys. D. Appl. Phys.* **39** 4563–7
[23] Sun Q, Di H, Li J and Wang X 2016 Effect of pulse frequency on microstructure and properties of welded joints for dual phase steel by pulsed laser welding *Mater. Des.* **105** 201–11
[24] Chmelickova H and Sebestova H 2012 Pulsed Laser Welding Nd Yag Laser ed D C Dumitras (Czech Republic: In Tech) p 41–58
[25] Kumar A, Young D and Ceglarek D 2013 Correlation analysis of the variation of weld seam and tensile strength in laser welding of galvanized steel *Opt. Lasers Eng.* **51** 1143–52
[26] Naderi M 2007 Hot Stamping of Ultra High Strength Steels (RWTH Aachen University)

IOP Conf. Series: Materials Science and Engineering **238** (2017) 012008 doi:10.1088/1757-899X/238/1/012008

Effect of shoulder to pin ratio on magnesium alloy Friction Stir Welding

N. H. Othman, M. Ishak and L. H. Shah

Manufacturing Process Focus Group, Faculty of Mechanical Engineering, Universiti Malaysia Pahang, 26600 Pekan, Pahang, Malaysia

E-mail : nurul_hidayah213@yahoo.com

Abstract. This study focuses on the effect of shoulder to pin diameter ratio on friction stir welding of magnesium alloy AZ31. Two pieces of AZ31 alloy with thickness of 2 mm were friction stir welded by using conventional milling machine. The shoulder to pin diameter ratio used in this experiment are 2.25, 2.5, 2.75, 3, 3.33, 3.66, 4.5, 5 and 5.5. The rotational speed and welding speed used in this study are 1000 rpm and 100 mm/min, respectively. Microstructure observation of welded area was studied by using optical microscope. Equiaxed grains were observed at the TMAZ and stir zone indicating fully plastic deformation. The grain size of stir zone increased with decreasing shoulder to pin ratio from ratio 3.33 to 5.5 due to higher heat input. It is observed that, surface galling and faying surface defect is produced when excessive heat input is applied. To evaluate the mechanical properties of this specimen, tensile test was used in this study. Shoulder to pin ratio 5.5 shows lowest tensile strength while shoulder to pin diameter ratio 3.33 shows highest tensile strength with weld efficiency 91 % from based metal.

1.Introduction

Weight reduction in automotive and aircraft industries is a main concern in improving fuel economy and reducing environmental pollutions [1].Recently, magnesium alloys are constantly gaining importance as lightweight structural materials for automotive applications [2]. The Mg alloys are especially attractive due to their low density, high specific stiffness and strength and also the recycling ability [3]. There have some problem in welding magnesium using conventional fusion welding. Conventional fusion welding methods for joining magnesium alloys produce some defects such as porosity and hot crack, which deteriorate their mechanical properties due to the melting material [4]. Friction stir welding (FSW) is a welding method under solid state joining that involves no melting, i.e. welding below melting point and it can eliminate problem related to solidification.[5]. The advantages of FSW compared to other method are low distortion and shrinkage even in a long weld, excellent mechanical properties in fatigue, tensile and bend test, no arc or fume, can weld in all position, no filler wire needed, weight reduction because no usage of filler, can weld aluminum and copper more than 75 mm with only one pass and uses consumable tools, which means that the tool can used for many times [6].

In FSW, the non-consumable tool plays an important role in generating heat by means of friction to plasticize the materials to be welded. FSW tool have three different portions such shank, shoulder and pin. The pin diameter and shoulder diameter decide the quantum of heat generated during FSW [7]. The tool shoulder diameter is having directly proportional relationship with the heat generation due to friction [8], [9]. The pin of the tool generates the heat and stirs the material being welded but the

IOP Conf. Series: Materials Science and Engineering **238** (2017) 012008 doi:10.1088/1757-899X/238/1/012008

shoulder also plays an important part by providing additional frictional treatment as well as preventing the plasticized material from escaping from the weld region [10].From the heating aspect, the relative size of pin and shoulder is important. The heat generation is related to the mechanical and microstructure properties of welding that affect the quality of the welding. Vijayavel et al. (2014) investigated the effect of five different shoulder diameters to pin diameter (D/d) ratio on tensile properties and hardness of the friction stir processed welding of metal matrix composite [11]. Another researcher, Noor Zaman Khan et al. (2015) also investigate the effect of D/d ratio on joint quality but done at AA6063 [12]. Both researcher used fix shoulder diameter for the experiment. Some question appear would the result same if use vary shoulder and pin diameter for the ratio and applied at AZ31B magnesium alloy. It was found that not many studies regarding the ratio of tool shoulder diameter with pin diameter on tensile properties, macrostructure, microstructures and hardness of FSW joints of AZ31B with varying both shoulder and pin diameter. This study investigates the effect of shoulder to pin ratio on magnesium alloy AZ31 friction stir welding.

2. Experimental Procedure

Experiment was performed on plate of AZ31 magnesium alloy with 2 mm thickness. The dimensions of base metal plate were 50 mm (width) × 60 mm (length) and the plate was longitudinally butt welded using conventional milling machine. Table 1 shows the chemical compositions of AZ31 alloy that was checked using Foundry Mass Spectrometers machine. The process parameters used for this experiment are given in Table 2 and the parameters are set from pre experiments, aiming at defect free specimens. The tilt angle was held constant at 3° based on several papers [13]–[16] and pre-tests. The welding tool material used is H13 steel with a taper cylindrical pin shape. The tool was processed by using a conventional lathe machine.

Table 1. Chemical compositions of AZ31 (wt.%).

Element	**Al**	**Zn**	**Mn**	**Si**	**Cu**	**Ca**	**Ni**	**Mg**
AZ31	2.87	0.99	0.239	0.0137	0.0183	0.0026	0.0015	Balance

Table 2. Welding parameters and tool dimensions for FSW.

Process parameters	Values
Rotational speed(rpm)	1000
Welding speed (mm/min)	100
Pin length, l (mm)	1.8
Tool shoulder diameter, D (mm)	9, 10, 11
Pin diameter, d(mm)	2, 3, 4
Pin taper angle	10°
Tool ratio, D/d	2.5, 2.75, 3, 3.33, 3.66, 4.5, 5 and 5.5
Holder diameter, a(mm)	20
Holder length, b(mm)	20
Shoulder length, c(mm)	24

Mechanical characterizations were conducted using tensile test. The tensile test was prepared according to the ASTM-E8 standard. The tensile test was conducted with three repetitions and the average values were then presented in result and discussion. The data were recorded and analyzed accordingly. Cross-sectional samples were then cut and mounted to analyze their microstructures. The mounted sample was ground manually using 240, 320, 400 and 600-paper grid, successively. After the grinding process, the specimens were polished using 1, 3 and 6 μm DIAMAT Polycrystalline diamond to remove the major scratches and finally etched using 4.2 g picric acid, 10 ml acetic acid, 10 ml H2O and 70 ml ethanol. The microstructure of the weld specimens was observed using an optical microscope.

IOP Conf. Series: Materials Science and Engineering **238** (2017) 012008 doi:10.1088/1757-899X/238/1/012008

3. Results and Discussion

3.1. Weld appearance

Visual inspection was used to characterize the weld bead defect and appearance. From figure below, observed that Figure 1 (i)-(m) shows smooth weld appearance while (a) ,(b) and (g) shows small flash at retreating side. Specimen with pin ratio 4.5, 5, and 5.5 shows flash defect. Most experiments that have flashed come from large tool ratio with small pin area that is 2 mm pin diameter. This is due to the higher heat generation at the weld center. This statement also support by Saravan et al. 2016 that shoulder give more friction resulting in higher heat generation while small pin diameter affect the splitting microstructure into fine grain size [17]. Experiment no 2 shows surface galling defect due to the sticking material at the pin [18] and ribbon flash is due to the over dwell time. When a longer time is taken after the initial plunge of the tool pin, it induces more heat to the specimen and increases the fluidity. The extra fluid (metal) will split out of the weld area, causing this defect [19]. All the defect will degrade the mechanical properties [18].

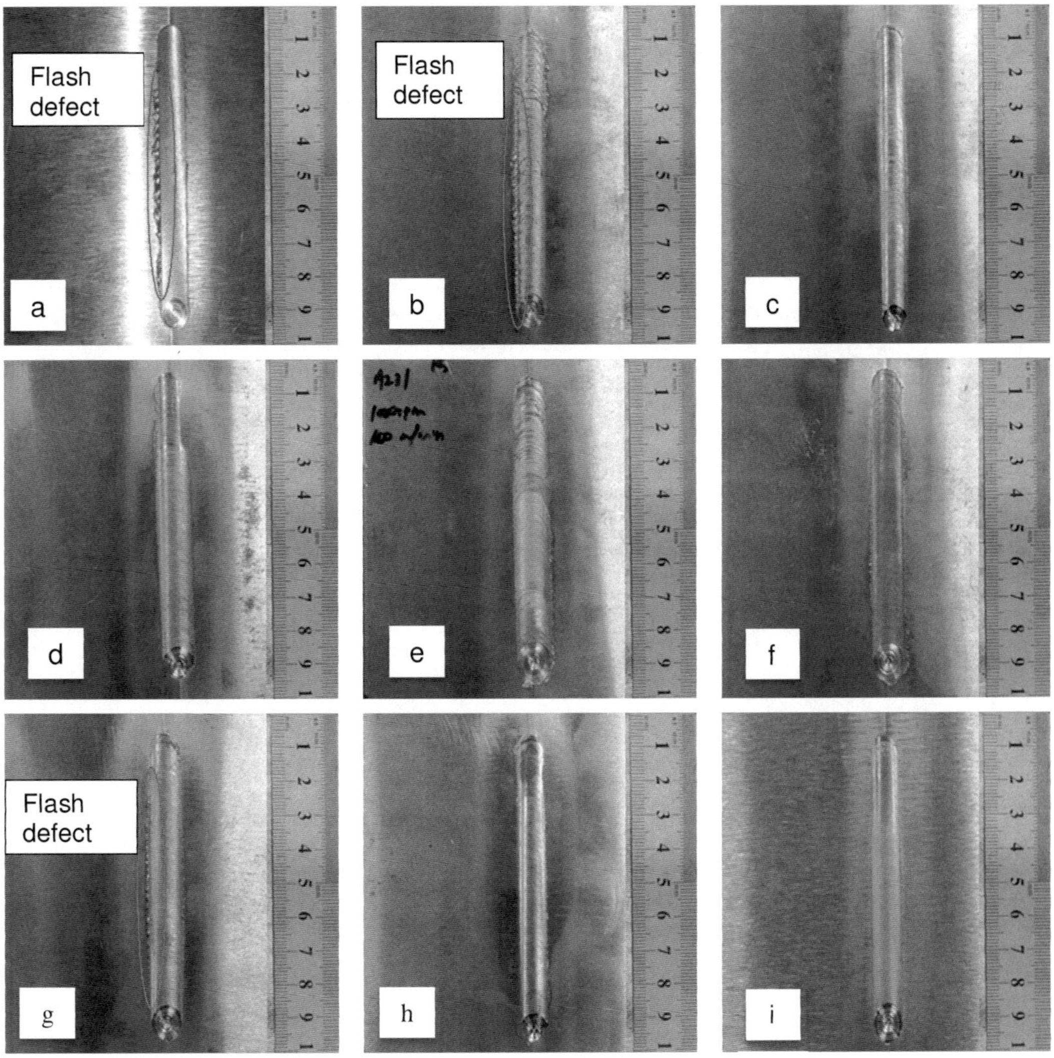

IOP Conf. Series: Materials Science and Engineering **238** (2017) 012008 doi:10.1088/1757-899X/238/1/012008

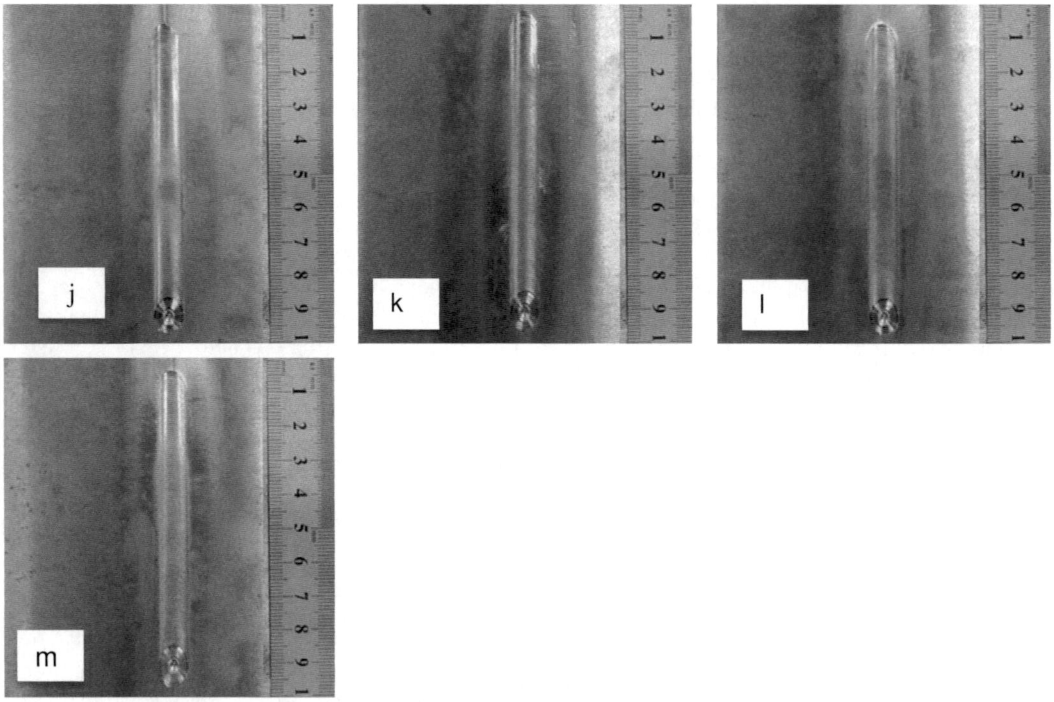

Figure 1. Weld appearances for shoulder to pin diameter ratio D/d on friction stir welding. a) ratio 4.5 , b) ratio 5.5 ,c) ratio 2.25, d) ratio 2.75 ,e) ratio 3 ,f) ratio 3.66 ,g) ratio 5, h) ratio 2.5 , i) ratio 3.33 ,j) ratio 3.33 , k) ratio 3.33 ,l) ratio 3.33 ,m) ratio 3.33.

3.2. Tensile properties of welded joint

The tensile test was conducted and the ultimate tensile strength (UTS) data shown in Table 3. The median from three samples for each specimen was taken to avoid the outliers from the overall samples.

Table 3. Ultimate tensile strength for shoulder to pin ratio friction stir welding.

Exp. No.	**Shoulder Diameter**	**Pin Diameter**	**Tool Ratio**	**Ultimate tensile strength (MPa)**
1	9	2	4.50	164.236
2	10	2	5.5	158.105
3	9	4	2.25	194.275
4	11	4	2.75	192.078
5	9	3	3	221.674
6	11	3	3.66	219.399
7	10	2	5	179.350
8	10	4	2.5	212.413
9	10	3	3.33	241.391
10	10	3	3.33	236.019
11	10	3	3.33	241.149

12	10	3	3.33	239.630
13	10	3	3.33	237.347

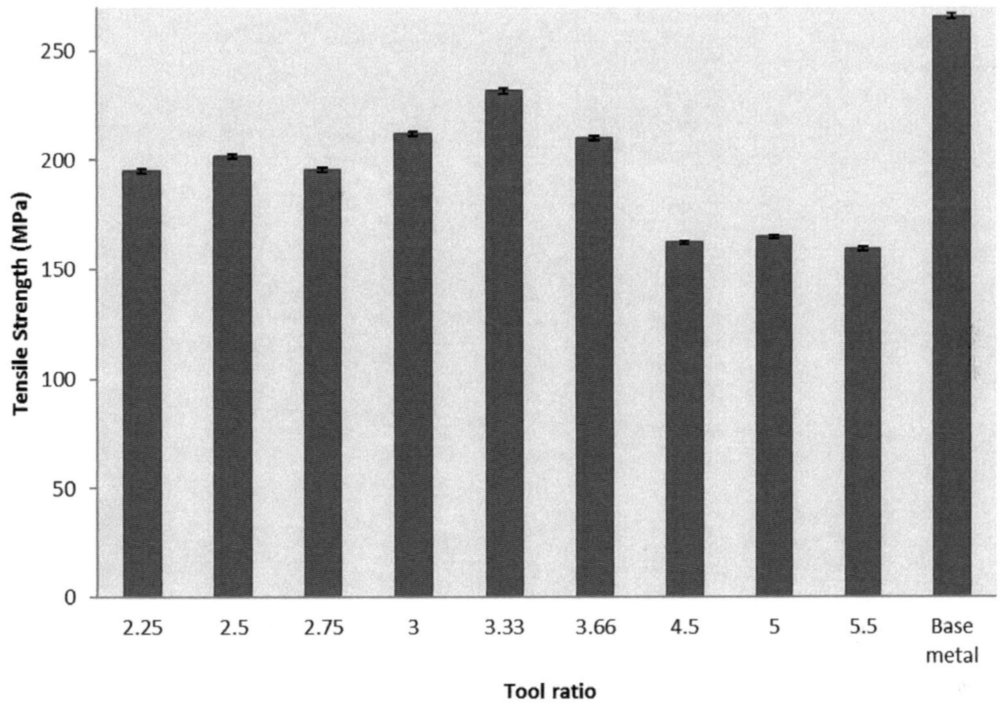

Figure 2. Ultimate tensile strength of different pin ratio.

According to the table, specimen 2 and 9 show the lowest and the highest UTS, respectively. The shoulder to pin ratio (D/d) 5.5 with shoulder diameter 11 and pin diameter 2 resulted in the lowest UTS value of 158.105 MPa. The shoulder to pin ratio (D/d) 3.3 with shoulder diameter 10 and pin diameter 3 resulted in the highest UTS value of 241.391 MPa. This value is near to the base metal UTS which is 265.6 MPa with 91 % weld efficiency.

Figure 2 shows the UTS of the welding joint and the base metals. The tensile strength trend shows that the strength are increase with increasing the tool ratio until tool ratio 3.33 and it decrease with increasing tool ratio after tool ratio 3.33. Tool ratio 4.5, 5 and 5.5 shows low tensile strength due to the flash defect shows in Figure 1. Flash may occur due to the improper parameter setting [20] that cause thinning the weld and result in decreasing the mechanical properties [18].

3.3. Metallurgical analysis

The weldability of magnesium alloy by using fiction stir welding was achieved through the excellent mechanical properties. Besides the mechanical properties, macro and microstructure of the welded area are also an important characteristic in identifying the ability of the material to form a good weld.

The sample for macrostructure observation was taken based on the tensile strength results; the lowest UTS and the highest UTS. The microstructural image was taken by using an optical microscope with ×50 magnifications. The cross sections of the welded area for lowest and highest tensile sample are shown in Figure 4 and Figure 6, respectively. It was observed that all specimens show the complete penetration. However, there were several defects that can be identified through the cross

IOP Conf. Series: Materials Science and Engineering **238** (2017) 012008 doi:10.1088/1757-899X/238/1/012008

section of the completed penetration specimen such as surface galling of weld surface and faying surface defect, as well as the porosity existent in the welded area as shown in Figure 3. Excessive hot weld is one of the factors existing surface galling and faying surface defect. Thus, the proper combination shoulder to pin diameter ratio are important to make sure the heat generation are sufficient for the welding.

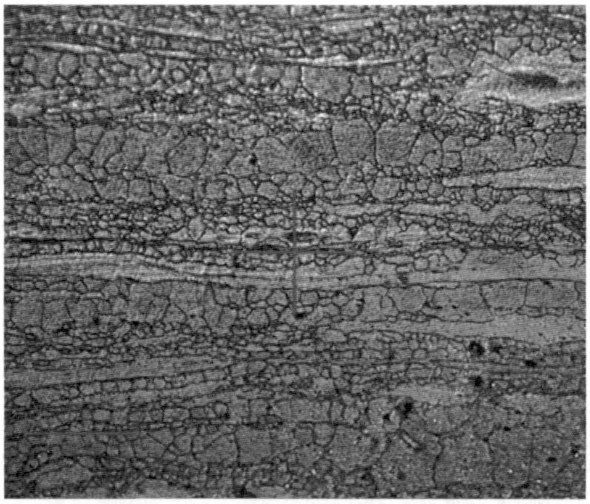

Figure 3.Microstructure of based metal (BM) AZ31 Magnesium alloy.

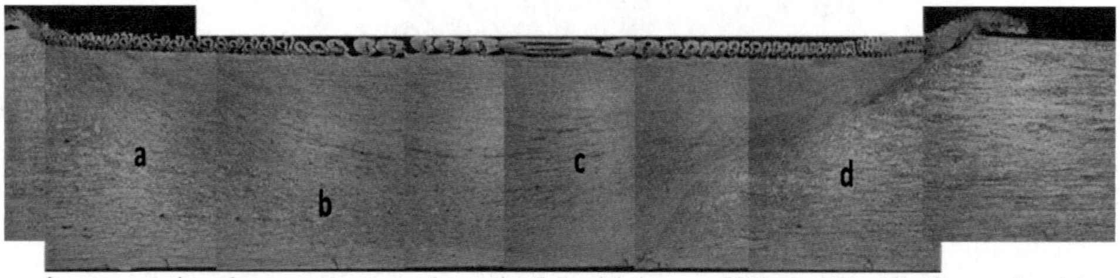

Figure 4.cross sectional macrostructure for ratio 5.5 with lowest Ultimate Tensile Strength. a) location retreating side HAZ, b) location TMAZ , c) location SZ , d) location advancing side HAZ.

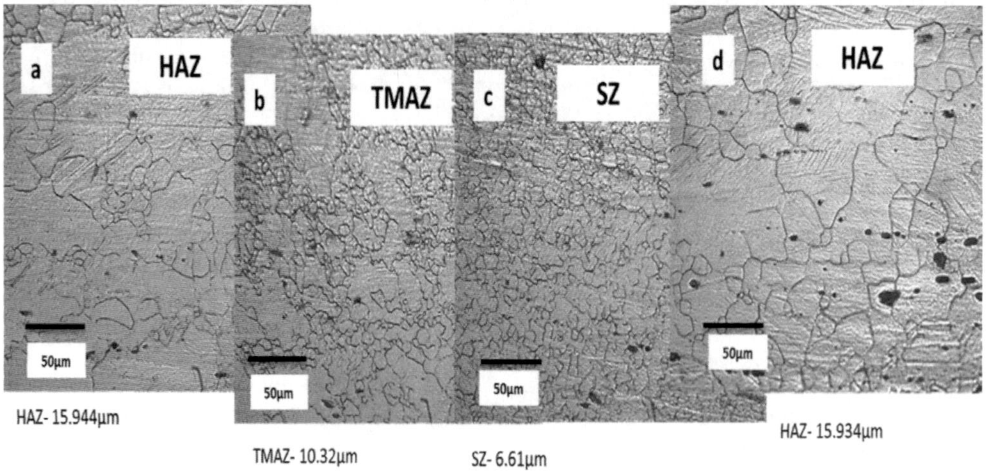

Figure 5. Microstructure of ratio 5.5 (a) Retreating side HAZ (b) TMAZ (c) SZ and (d) Advancing side HAZ.

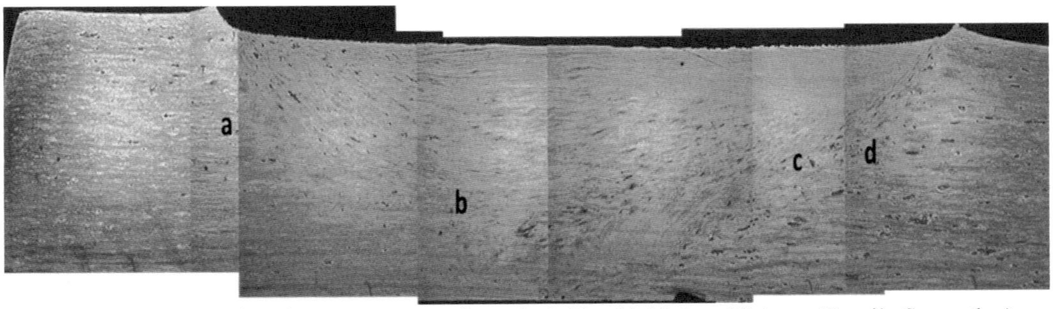

Figure 6. Cross sectional macrostructure for ratio 3.33 with highest Ultimate Tensile Strength a) location retreating side HAZ, b) location SZ ,c) location TMAZ, d) location advancing side HAZ.

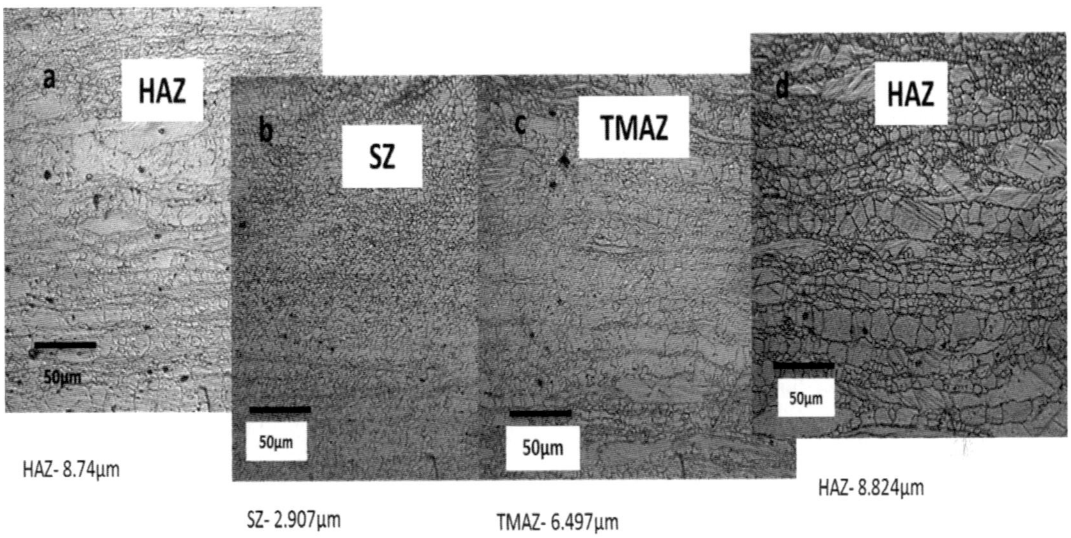

Figure 7. Microstructure of ratio 3.33 (a) Retreating side HAZ (b) TMAZ (c) SZ and (d) Advancing side HAZ.

The major FSW regions can be observed, namely the base metal (BM), heat affected zone (HAZ), thermo mechanically affected zone (TMAZ) and weld zone (WZ). Figure 3 shows typical microstructure of base metal (BM) before friction stir weld. The base metal exhibit coarse microstructure with elongated and pancake grain size. Figure 5 (b) and (c) shows the TMAZ and SZ for the lowest tensile strength. The grain size at SZ shows fine grain size. TMAZ shows elongated grain compare to SZ with average grain size 10.32 μm for TMAZ and 6.61 μm for SZ. HAZ shows uneven grain boundary with average grain size 15.944 μm. Shoulder give more friction resulting in higher heat generation but the small pin diameter affect the splitting microstructure into fine grain size [17]. Figure 6 shows cross sectional microstructure for ratio 3.33 with highest tensile strength. SZ and TMAZ at ratio 3.33 consist of fine and equiaxed grains which show that recrystallization already occur. Small grain size affect the tensile strength due to the homogeneous in whole area and crystal orientations of such grains are rather randomly distributed [21]. TMAZ shows elongated grain compare to SZ with average grain size 6.497 μm and 2.907 μm respectively. HAZ shows coarser and elongated grain size compare to SZ and TMAZ but small HAZ grain size compare to tool ratio 5.5.

The stir zone exhibit fine grain compared to base metal which is coarse grain. This behavior is due to dynamic recrystallization from the simultaneous deformation and heat input during welding process where the heat generated from the interaction of the pin tool deformed the grain boundaries [22] lead to the higher tensile strength of specimen. Grain structures for TMAZ in Figure 7 have slightly changed in grain structure due to the heat input that is sufficiently heated to undergo deformation from its original grain boundaries. TMAZ region undergo plastic deformation but recrystallization did not take place in this region due to inadequate deformation strain [23]. In HAZ, the magnesium alloy experienced thermal cycle but only little plastic deformation occurred. The increase in grain size consequently decreased the tensile strength [24]. Smaller grain size will affect the tensile strength due to the dislocation movement in the material. By reducing the dislocation movement of grain (fine grain size), the mechanical strength of material can be improve due to the metal plastically hard to deform [25].

4. Conclusion

1. Thickness of 2 mm magnesium alloy AZ31 plates were successfully welded by using friction stir welding process with shoulder to pin diameter ratio 2.5, 2.75, 3, 3.33, 3.66, 4.5, 5 and 5.5 with 1000 rpm rotational speed and 100 mm/min welding speed
2. All specimen produce sound quality joints with minimal defects. The best weld appearance can be observed in small ratio from 2.25 till ratio 3.33. Ratio 3.66 until 5.5 shows some defect such as little flash defect. Fine grain size can be seen in WZ and TMAZ and coarser grain size can be seen in the HAZ. The average grain size for D/d ratio 3.33 shows smaller grain size compare to other tool ratio. Smaller grain size will affect the tensile strength of the specimens. The D/d tool ratio 3.33 yielded the highest tensile value of 241.391 MPa due to the fine grain size and good weld appearance. Therefore the optimum D/d shoulder to pin ratio for magnesium alloy AZ31 are at ratio 3.33 gives better weld appearance and high tensile strength.

Acknowledgment

The author would like to thank the supervisors and technical staffs in Universiti Malaysia Pahang, all of the work within which experiment was conducted. The financial support by Ministry of Education Malaysia through Universiti Malaysia Pahang for Research Grant project no. RDU140118 is also grateful acknowledged.

References

[1] Cao X and Jahazi M 2009 Effect of welding speed on the quality of friction stir welded butt joints of a magnesium alloy *Mater. Des.* **30** (6) 2033–2042

[2] Sahu G A 2015 Brief Review on MG Alloys Their Properties and Application *Int. J. Adv. Res. Sci. Eng.* 8354 (4) 65–71

[3] Blawert C, Hort N and Kainer K U 2006 Automotive Applications of Magnesium and Its Alloys *Magnesium* 57 (4) 397–408

[4] Rajakumar S, Razalrose A and Balasubramanian V 2013 Friction stir welding of AZ61A magnesium alloy: A parametric study *Int. J. Adv. Manuf. Technol.* 68 (1–4) 277–292

[5] Mishra R S and Ma Z Y 2005 Friction stir welding and processing *Mater. Sci. Eng. R Reports* 50 (1–2) 1–78

[6] Thomas W M U, Nicholas E D, Hall A and Cb C 1998 Friction stir welding for the transportation industries 18 269–273

[7] Malarvizhi S and Balasubramanian V 2012 Influences of tool shoulder diameter to plate thickness ratio (D/T) on stir zone formation and tensile properties of friction stir welded dissimilar joints of AA6061 aluminum–AZ31B magnesium alloys *Mater. Des.* 40 453–460

[8] Reddy G M, Mastanaiah P, Murthy C V S and Mohandas T 2006 Microstructure , Residual Stress Distribution and Mechanical Properties of Friction-Stir AA 6061 Aluminium Alloy Weldments *Proc. Nat. Seminar on Non-Destructive Evaluation*

[9] Oosterkamp A A N A and Djapic Oosterkamp L 2004 Kissing Bond Phenomena in Solid-State Welds of Aluminum Alloys 225–231

[10] Rajakumar S, Muralidharan C and Balasubramanian V 2011 Influence of friction stir welding process and tool parameters on strength properties of AA7075-T6 aluminium alloy joints *Mater. Des.* **32** (2) 535–549
[11] Vijayavel P, Balasubramanian V and Sundaram S 2014 Effect of shoulder diameter to pin diameter (D/d) ratio on tensile strength and ductility of friction stir processed LM25AA-5% SiCp metal matrix composites *Mater. Des.* **57** 1–9
[12] Khan N Z, Khan Z A and Siddiquee A N 2015 Effect of Shoulder Diameter to Pin Diameter (D/d) Ratio on Tensile Strength of Friction Stir Welded 6063 Aluminium Alloy *Mater. Today Proc.***2** (4–5) 1450–1457
[13] Chen Y C and Nakata K 2008 Friction stir lap joining aluminum and magnesium alloys *Scr. Mater.* **58** (6) 433–436
[14] Silva A A M, Arruti E, Janeiro G, Aldanondo E, Alvarez P and Echeverria A 2011 Material flow and mechanical behaviour of dissimilar AA2024-T3 and AA7075-T6 aluminium alloys friction stir welds *Mater. Des.* **32** (4) 2021–2027
[15] Othman N H, Shah L H and Ishak M 2015 Mechanical and microstructural characterization of single and double pass Aluminum AA6061 friction stir weld joints *IOP Conf. Ser. Mater. Sci. Eng.***100** 012016
[16] Sathari N A A, Razali A R, Ishak M and Shah L H 2015 Mechanical Strength of Dissimilar Aa7075 and Aa6061 Aluminum Alloys Using Friction Stir Welding *Int. J. Automot. Mech. Eng.* **11** 2713–2721
[17] Saravanan V, Rajakumar S, Banerjee N and Amuthakkannan R 2016 Effect of shoulder diameter to pin diameter ratio on microstructure and mechanical properties of dissimilar friction stir welded AA2024-T6 and AA7075-T6 aluminum alloy joints *Int. J. Adv. Manuf. Technol.*
[18] Arbegast W J 2003 Friction Stir Joining : Characteristic Defects *Adv. Mater. Process. Cent. MET* **6** 1–30
[19] Babu A S and Devanathan C 2013 An Overview of Friction Stir Welding *Int. J. Res. Mech. Eng. Technol.* **3** (2) 59–265
[20] Gibson B T, Lammlein D H, Prater T J, Longhurst W R, Cox C D, Ballun M C, Dharmaraj K J, Cook G E and Strauss A M 2014 Friction stir welding: Process, automation, and control *J. Manuf. Process.* **16** (1) 1–18
[21] Goloborodko A, Ito T, Yun X, Motohashi Y and Itoh G 2004 Friction Stir Welding of a Commercial 7075-T6 Aluminum Alloy: Grain Refinement, Thermal Stability and Tensile Properties *Mater. Trans.* **45** (8) 2503–2508
[22] Dawood H I, Mohammed K S and Rajab M Y 2014 Advantages of the Green Solid State FSW over the Conventional GMAW Process *Adv. Mater. Sci. Eng.* **2014** 1–10
[23] Lohwasser D and Chen Z 2010 *Friction stir welding: From basics to applications.* (Cambridge, England: Woodhead Publishing Limited)
[24] Rajakumar S, Muralidharan C and Balasubramanian V 2010 Establishing empirical relationships to predict grain size and tensile strength of friction stir welded AA 6061-T6 aluminium alloy joints *Trans. Nonferrous Met. Soc. China* **20** (10) 863–1872
[25] Abbaschian R, Abbaschian L and Reed-Hill R E 2009 *Physical Metallurgy Principles.* (Cengage)

IOP Conf. Series: Materials Science and Engineering **238** (2017) 012009 doi:10.1088/1757-899X/238/1/012009

Investigation on the Effect of Pulsed Energy on Strength of Fillet Lap Laser Welded AZ31B Magnesium Alloys

M N M Salleh[1], M Ishak[1,2], M H Aiman[1], S R A Idris[1] and F R M Romlay[1,2]

[1]Faculty of Mechanical Engineering, Universiti Malaysia Pahang, 26600 Pekan, Pahang
[2]Automotive Engineering Centre (AEC), UMP, 26600 Pekan, Pahang

E-mail: mdnaqiuddinsalleh91@gmail.com

Abstract. AZ31B magnesium alloy have been hugely applied in the aerospace, automotive, and electronic industries. However, welding thin sheet AZ31B was challenging due to its properties which is easily to evaporated especially using conventional fusion welding method such as metal inert gas (MIG). Laser could be applied to weld this metal since it produces lower heat input. The application of fiber laser welding has been widely since this type of laser could produce better welding product especially in the automotive sectors. Low power fiber laser was used to weld this non-ferrous metal where pulse wave (PW) mode was used. Double fillet lap joint was applied to weld as thin as 0.6 mm thick of AZ31B and the effect of pulsed energy on the strength was studied. Bond width, throat length, and penetration depth also was studied related to the pulsed energy which effecting the joint. Higher pulsed energy contributes to the higher fracture load with angle of irradiation lower than 3 °

1. Introduction

Laser welding applications have been widely used in this century associated with the joining technology developments. There were a few commonly used lasers welding in industry such as CO2 gas laser, Nd: YAG laser, Disk Laser, and Fiber laser. Fiber laser is the newest laser technology where it was claimed to be the best method among the other laser welding types especially in term of operating cost [1]. In laser welding application, materials such as ferrous and non-ferrous metal could be joint it was proven that this method is a new welding technique which can produce high power with lower heat input, high strength joint and low distortion [2, 3]. A few of researcher have been used laser welding to joint non-ferrous alloys such as aluminium and magnesium alloys which is difficult to be joint due to their lower melting point compared to steel. Fiber laser is one of Q-Switch type lasers which could generate two modes, which is pulse wave (PW) mode and continuous wave (CW) mode. In this paper, PW mode was selected because it also could produce continuous welding with the usage of lower laser power compared to CW mode. Different values of pulsed energy will be studied in this experiment where the calculation of pulsed energy as shown in Equation (1) below.

$$\text{Average Peak Power (W)} = (\text{Pulsed Energy (EP)})/(\text{Pulse Duration }(\Delta t)) \quad (1)$$

Past researcher used PW mode in welding application of steels, alloys, and thin sheets metal [4-7]. However, they neglected the effect of pulsed energy on the welded metal. AZ31B was selected in this experimental work due to its characteristics of super light density, and high-resistance of corrosion. It was firmly used in automobile industry as it can reduce the cost and weight of the product [8, 9]. So far, the study on laser welding of AZ31B Magnesium alloys has been reduced especially towards the thin sheets AZ31B. Usually, unstable weld pool and crack was found on post weld magnesium. However, AZ31B could be welded with crack-free joints with lesser porosity and good surface quality [10, 11]. The intention of this experimental work is to investigate the effect of pulsed energy, EP of PW mode by low power fiber laser on the macrostructure evaluation and the joint strength.

2. Experimental Procedure

2.1. Material

AZ31B Magnesium alloy with thickness of 0.6 mm was used in this study. The major alloying element for this material is aluminium (Al), and zinc (Zn). Before the metal subjected to the laser welding, it was cut by dimension 53 × 20 × 0.6 mm. Samples were polished with 180 grit sand paper in order to remove the oxide layer presented at the welding area. The element composition was presented in Table 1 which collected from the spectrometer test.

Table 1. Chemical composition for as received AZ31B.

Element	Al	Zn	Mn	Si	Cu	Fe	Mg
Weight,%	2.8	1.2	0.4	0.02	0.002	0.02	Bal.

2.2. Welding Setup

Pulse wave (PW) mode was used in this experiment where the pulsed energy, EP used are 1.8, 2.0, and 2.2 J with constant pulse width and pulse repetition rate of 2 ms and 60 HZ, respectively. The angle of irradiations used is 2, 3, and 4 degrees and the welding speed used is 2, 3, and 4 mm/s. Figure 1 shows the laser welding condition in order to produce double fillet lap joint.

Fifteen samples were welded in this experimental work and the bond width, throat length, and penetration depth of all samples were measured and collected. Figure 2 shows the schematic illustration of the macrostructure evaluation. Tensile-shear test was carried out for all welded samples. The average values of three tested samples were collected accordingly to the number of experiment.

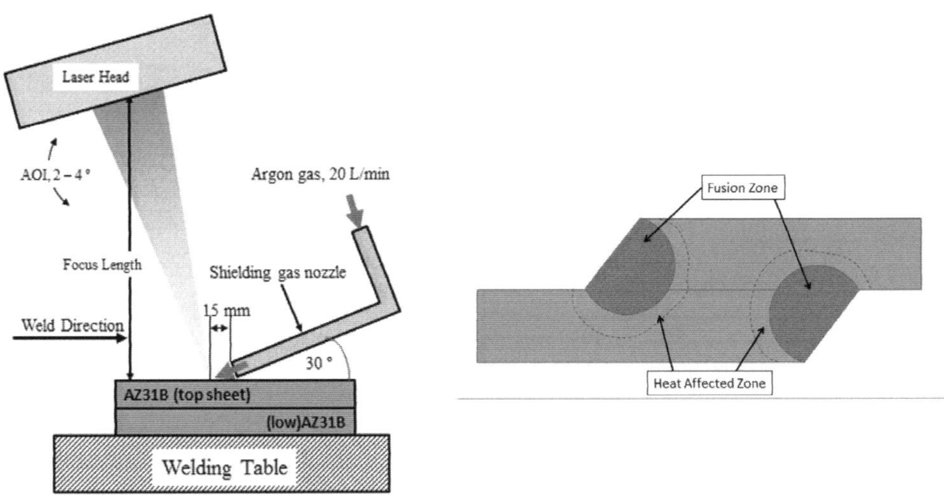

Figure 1. Double fillet lap laser welding condition.

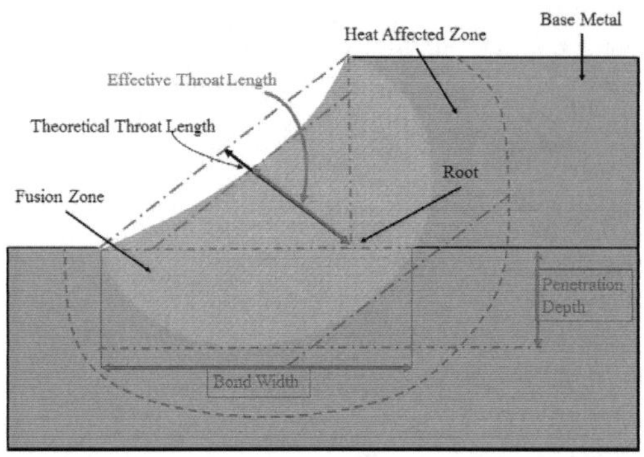

Figure 2. Schematic illustration of macrostructure evaluation.

3. Results and Discussion

The values of bond width and throat length were calculated and tabulated in Table 2 below. Bond width is the highest factor that determines the strength of the joint instead of penetration depth in case of lap joint method. The effects of the parameters on the fracture load were analyzed based on the macrostructure evaluation where the bond width, throat length, and penetration depth are measured and tabulated in Table 3.

Table 2. Fracture load result according to experiment order.

Sample	EP (J)	WS (mm/s)	AOI (°)	Fracture Load (N)
1	1.8	2	3	101.5
2	2.2	2	3	589.8
3	1.8	4	3	195.5
4	2.2	4	3	482.7
5	1.8	3	2	257.6
6	2.2	3	2	629.3
7	1.8	3	4	515.5
8	2.2	3	4	549.1
9	2.0	2	2	740.2
10	2.0	4	2	593.4
11	2.0	2	4	292.1
12	2.0	4	4	180.4
13	2.0	3	3	520.0
14	2.0	3	3	426.3
15	2.0	3	3	450.0

From Table 2, it was observed that sample 9 welded by pulsed energy 2.0 J, 2 mm/s welding speed and 2 ° irradiation angle has the highest fracture load with a value of 740.2 N. Meanwhile, the lowest fracture load possessed by sample 1 (EP 1.8 J, WS 2 mm/s, 3°) with a value of 101.5 N which was 7.29 % lower than the highest one. It was observed from this table that lower pulse energy with higher angle of irradiation contributes to the lower joint strength. From Table 3, it was observed that samples which have undercut defects experienced the lower fracture load. It was clearly seen that there was no undercut measured from the cross section of sample 9 compared to sample 1 which has 31 μm undercut depth. Highest undercut depth was observed from sample 2 (151 μm), however the fracture load still higher since it was welded with pulsed energy 2.2 J. From Table 3, three graphs were constructed in order to observe the effect of different pulse energy to fracture load when it affects the results of bond width, actual throat length, and penetration depth of the double fillet lap product.

Table 3. Fracture load result according to experiment order.

Sample	Bond Width (μm)	Actual Throat Length (μm)	Penetration Depth (μm)	Undercut (μm)
1	734	241	353	31
2	1079	156	802	151
3	669	199	270	43
4	1026	197	810	209
5	736	208	523	70
6	610	342	217	None
7	883	244	566	50
8	884	182	779	70
9	806	236	357	None
10	581	285	109	None
11	1090	232	629	125
12	736	204	383	58
13	674	196	347	None
14	664	208	346	None
15	670	195	342	None

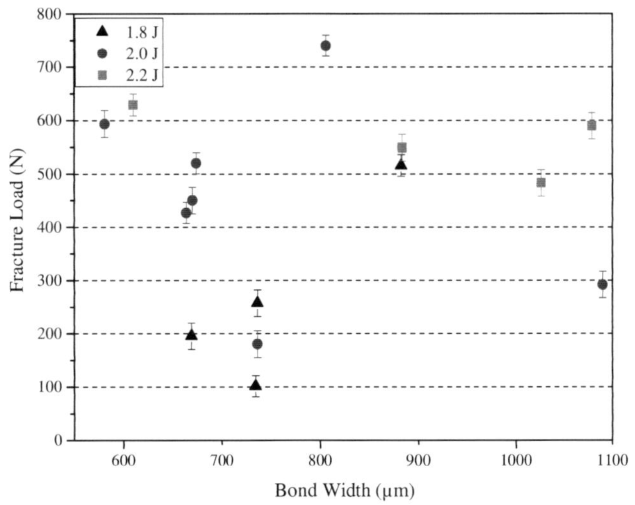

Figure 3. Fracture Load (N) versus bond width (μm) at different pulsed energy.

Figure 3 shows the relationship between average value of fracture load and the bond width of the joint at varying pulsed energies. High fracture load could be achieved as the pulsed energy is applied at 2.0 J with the middle value of bond width. The highest fracture load was observed from sample 9 with fracture load of 740 N which had 806 μm of bond width with no undercut defect presented as shown in Table 3. However, the fracture load was found to be lower at bond width of 736 μm although 2.0 J energy was used due to the undercut defect presents which has a depth of 58 μm. Larger bond width (1090 μm = 10.9 mm) produced by a 2.0 J pulse energy creates as low as 292.1 N as observed from sample 11.

High undercut depth of 125 μm was measured which was thought to be the main factor to the lower joint strength. Based on the figure, the lowest fracture load was found at the bond width 734 μm possessed by sample number 1 where the pulse energy used is 1.8 J. Undercut defect was observed from sample 1 with 31 µm depth. It has been observed that sample welded by pulsed energy of 2.2 J (highest energy used in experiment) produced the fracture load higher than 450 N. The highest fracture load as using pulse energy of 2.2 J was observed at 629 N and the lowest fracture load at 482.7 N as shown in Table 2.

Figure 4 and 5 show the relationship of the throat length and fracture load to the fracture load with three types of pulsed energy values. From Figure 4, it was shown that the highest fracture load was possessed by the samples with the pulse energy of 2.0 J and the throat length of 240 µm. At the same throat length of 240 µm with pulsed energy of 1.8 J, the fracture load was dropped to approximately 100 N. With throat length of 156 and 342 µm (sample 2 and 6), the fracture load was found to be approximately same at 600 N with the same pulsed energy of 2.2 J.

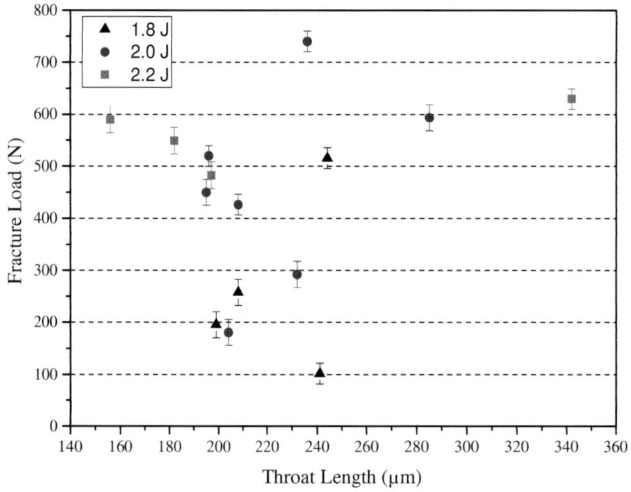

Figure 4. Fracture Load (N) versus throat length (µm) at different pulsed energy.

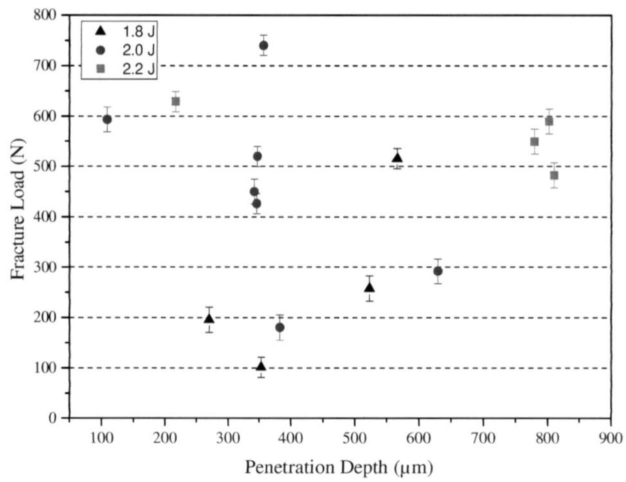

Figure 5. Fracture Load (N) versus penetration depth (µm) at different pulsed energy.

However, the differences in welding speed and angle of irradiation contributed to the larger throat length. Based on the figure, it was observed that the throat length increased to above 260 µm as the pulse energy used was 2.0 and 2.2 J. Meanwhile, pulse energy of 1.8 J produced throat length less than

260 μm with fracture load less than 600 N. Higher fracture load above 600 N could be achieved as the pulsed energy used above than 2.0 J with throat length 240 until 340 μm large. For Figure 5, it was observed that lower penetration depth of 109 μm (sample 10) possessed almost 600 N of fracture load using pulsed energy 2.0 J. However, the fracture load dropped to less than 300 N using 2.0 J of pulsed energy although the penetration depth was increased to 400 until 650 μm (sample 11 and 12).

It was observed in Table 3 that sample 11 and 12 possessed undercut defect at the weld region with undercut depth of 125 and 58 μm, respectively. For sample welded by higher pulsed energy (2.2 J), the penetration depth achieved was approximately 800 μm which indicated that the samples have already produced excess weld penetration. The fracture load produced for these samples were less than 600 N. However, higher fracture load above 600 N using the same pulsed energy could be achieved with penetration depth of 217 μm which was half of base metal thickness. Fracture load less than 600 N possessed by all samples welded with 1.8 J of energy.

4. Conclusion

The relationship of the fracture load with the macrostructure appearances such as bond width, throat length, and penetration depth with different values of pulsed energy has been studied. Fracture load as high as 740.2 N has been achieved with the bond width, throat length, and penetration depth of 806, 236, and 357 μm, respectively. The effective pulsed energy, EP to produce higher fracture load was 2.0 J. Based on the experiment, it could be concluded that the joint strength was higher if the usage of pulsed energy was higher as it could produce larger bond width. In addition, the usage of angle of irradiation affects the joint strength since it could produce undercut defect. Undercut defect could contributes to the lower joint strength.

Acknowledgement

The work is supported under grant GRS150345 for the financial. All experimental works were done in Joining and Welding lab, Faculty of Mechanical Engineering, UMP

References

[1] Assunção E, Quintino L and Miranda R 2009 Comparative study of laser welding in tailor blanks for the automotive industry *The International Journal of Advanced Manufacturing Technology* **49** 123-131

[2] Kenji O and Masatoshi M 2015 *Recent Trend of Welding Technology Development and Applications* (JFE Holdings, Inc, Chiyodaku, Tokyo, Japan)

[3] Yan F, Liu S, Hu C et al. 2017 Liquation cracking behavior and control in the heat affected zone of GH909 alloy during Nd: YAG laser welding *Journal of Materials Processing Technology* **244** 44-50

[4] Esraa K H, Farah H and Makram F 2014 Laser wavelength and energy effect on optical and structure properties for nano titanium oxide prepared by pulsed laser deposition *Iraqi Journal of Physics* **12** 62-68

[5] Ishak M, Yamasaki K, Maekawa K 2009 Lap Fillet Welding of Thin Sheet AZ31 Magnesium Alloy with Pulsed Nd:YAG Laser, *Journal of Solid Mechanics and Materials Engineering* **3** 1045-1056

[6] Levin Y Y and Erofeev V A 2009 Calculation of the parameters of pulsed laser welding of thin sheets of aluminium alloys *Welding International* **23** 934-938

[7] Ventrella V A, Berretta J R and de Rossi W 2011 Micro Welding of Ni-based Alloy Monel 400 Thin Foil by Pulsed Nd:YAG laser *Physics Procedia* **12** 347-354

[8] Liu Q, Zhou X, Zhou H, et al. 2017 The effect of extrusion conditions on the properties and textures of AZ31B alloy *Journal of Magnesium and Alloys*

[9] Brandizzi M, Renna S, Satriano A, et al. 2013 Nd:YAG laser welding of fine sheet metal butt joints in AZ31 magnesium alloy *Welding International* **28** 784-792

[10] Harooni M, Ma J, Carlson B, et al. 2015 Two-pass laser welding of AZ31B magnesium alloy *Journal of Materials Processing Technology* **216** 114-122

[11] Casalino G, Guglielmi P, Lorusso V D, et al. 2017 Laser offset welding of AZ31B magnesium alloy to 316 stainless steel *Journal of Materials Processing Technology* **242** 49-59

IOP Conf. Series: Materials Science and Engineering **238** (2017) 012010 doi:10.1088/1757-899X/238/1/012010

Influence of multi-walled carbon nanotubes on melting temperature and microstructural evolution of Pb-free Sn-5Sb/Cu solder joint

T T Dele-Afolabi[1], M A Azmah Hanim[1,4], M Norkhairunnisa[2,4], M T Suraya[1] and H M Yusoff[3]

[1] Department of Mechanical and Manufacturing Engineering, Faculty of Engineering, Universiti Putra Malaysia, 43400 UPM Serdang Selangor, Malaysia.
[2] Department of Aerospace Engineering, Faculty of Engineering, Universiti Putra Malaysia, 43400 UPM Serdang Selangor, Malaysia.
[3] Department of Chemical and Environmental Engineering, Faculty of Engineering, Universiti Putra Malaysia, 43400 UPM Serdang Selangor, Malaysia.
[4] Laboratory of Biocomposite Technology, Institute of Tropical Forestry and Forest Products (INTROP), Universiti Putra Malaysia, 43400 UPM Serdang, Selangor Darul Ehsan, Malaysia

E-mail: azmah@upm.edu.my

Abstract. In this study, the effects of multi-walled carbon nanotubes on the melting temperature and microstructural evolution of the Sn-5Sb/Cu joints are evaluated. Plain and carbon nanotubes (CNTs) reinforced Sn-5Sb solder systems with solder formulations Sn-5Sb, Sn-5Sb-0.01CNT, Sn-5Sb-0.05CNT and Sn-5Sb-0.1CNT were prepared through the powder metallurgy route and thereafter samples were subjected to thermal and microstructural evaluation. As retrieved from the DSC scans, a slight decline in the peak temperature was observed in the composite solders which is indicative of the CNTs role in exciting surface instability in the host Sn matrix. In order to prepare the solder joints and analyze the interfacial intermetallic compound (IMC) evolution, respective solder systems were placed on copper (Cu) substrate and subjected to both reflow soldering and isothermal aging (170°C) conditions. From the IMC thickness result, considerable retardation in the IMC layer growth was observed in the CNTs reinforced solder joints, especially the 0.05wt.% CNTs solder system owing to the inhibition of Sn atoms diffusion by reinforcement material.

1. Introduction

Taking into consideration the ongoing miniaturization of electronic products, residual stress increase in interconnection joints is imminent owing to the space restriction and the thermal fluctuation experienced during the service life of the products. Meanwhile, due to a sharp rise in the input/output terminals during electronic packaging [1-3], achieving high reliability and maintaining an acceptable level of performance efficiency is a major point of call for researchers in the development of excellent lead-free solder alternatives.

As highlighted in the study of [4] alloying lead-free solders with small-scale content of foreign elements is a viable way of influencing the microstructure, the reaction growth rate and the physical properties of the phases formed. In recent times, nano-sized particles have been employed as reinforcement materials in lead free solders due to their effectiveness in restricting grain boundary sliding through uniform distribution of the particles at the grain boundaries [5,6]. A typical nano-

IOP Conf. Series: Materials Science and Engineering **238** (2017) 012010 doi:10.1088/1757-899X/238/1/012010

material that has caught the interest of researchers and has shown to be a promising reinforcement material is the carbon nanotubes (CNTs)

Consolidating on the advantages of CNTs, [7-9] reported that the Sn-Ag-Cu (SAC) solder system reinforced with single-walled carbon nanotubes (SWCNTs) exhibited a slight decline in the melting temperature as against that of the plain solder. Similarly, Han et al. fabricated a composite solder system through the incorporation of Ni-CNTs dispersoids [10]. The authors observed an insignificant alteration in the melting temperature of Ni-CNTs doped solders which ranged from 219.4°C to 220.8°C. Elsewhere, Mayappan et al. reported that the addition of CNTs inhibited the growth of Cu6Sn5 and Cu3Sn from 1.69 x 10-10 - 1.02 x 10-10cm2/s and 4.9 x 10-11- 2.5 x 10-11cm2/s respectively [11].

In the studies of Xu et al., the authors revealed the evolution of a retarded overall IMC (Cu_6Sn_5 and Cu_3Sn) layer growth in CNTs reinforced solder than those of the unreinforced counterpart after subjecting both solder systems to 336 h aging condition with a continuous current density of 1.2 x 104 A/cm2 [12]. So far, investigations [13,14] have shown the SAC solder to be the preferential candidate for lead-free soldering application hence instigating the neglect of other lead-free solder counterparts. Belonging to this category is the Sn-5Sb solder whose unique melting temperature of 240°C makes it a dual purpose solder suitable for step soldering and high temperature application in electronic packaging [15]. However, the solder experiences extreme IMC layer growth during its operational lifetime which poses great danger to the mechanical integrity of the joints developed [16].

In the present study, plain and composite Sn-5Sb solder systems were developed through the powder metallurgy method. The thermal properties of the developed solder systems and the intermetallics evolution of these solder systems on a Cu substrate were investigated. With the inclusion of the CNTs phase, it is envisioned that the developed joint will function decently in a robust operating environment.

2. Experimental Procedure

Multi walled carbon nanotubes (MWCNTs) having a diameter range of 15-20nm and a length range of 0.5-2μm (see Figure. 1), was used as the reinforcing material in this study. The Sn-5Sb, Sn-5Sb-0.01CNT, Sn-5Sb-0.05CNT and Sn-5Sb-0.1CNT solder samples were prepared by the ball milling process at 800rpm for 6h using a planetary mono mill (Fritsch Pulverisette 7). Cold compaction of the milled powders was performed on the universal testing machine (INSTRON 3382) using a pressure of 80MPa to form a 20mm solder pellet.

Differential scanning calorimetry (DSC) analysis was carried out using Mettler Toledo DSC 823e model to study the melting properties of the solder samples. Solder samples were heated up from 30°C till 600°C at 10°C/min heating rate. Thereafter, solder samples were reflowed on a pure (99.99%) Cu substrate with a thin layer flux application. After the reflow process, some of the samples were held back and subjected to solid-state isothermal aging for 500 hours.

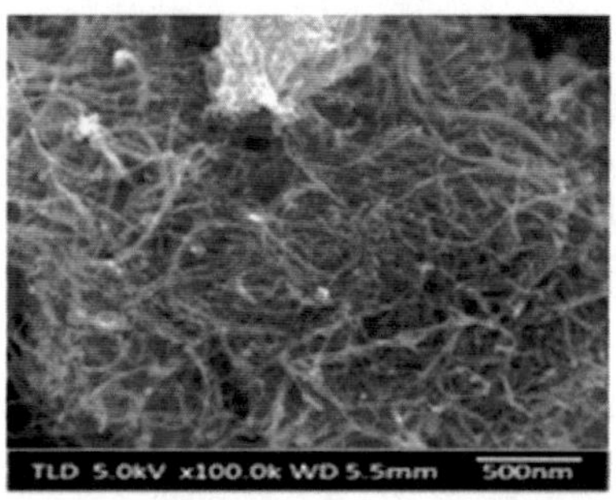

Figure 1. SEM image of MWCNTs.

3. Results and Discussion

3.1. Melting temperature analysis

Considering the surging technological advancement in electronics packaging, it is pertinent to comprehensively study the melting temperature properties of the solder interconnection materials. Figure 2 and Table 1 presents both the differential scanning calorimetry (DSC) scans and the melting temperature data for both the plain and composite solder systems respectively.

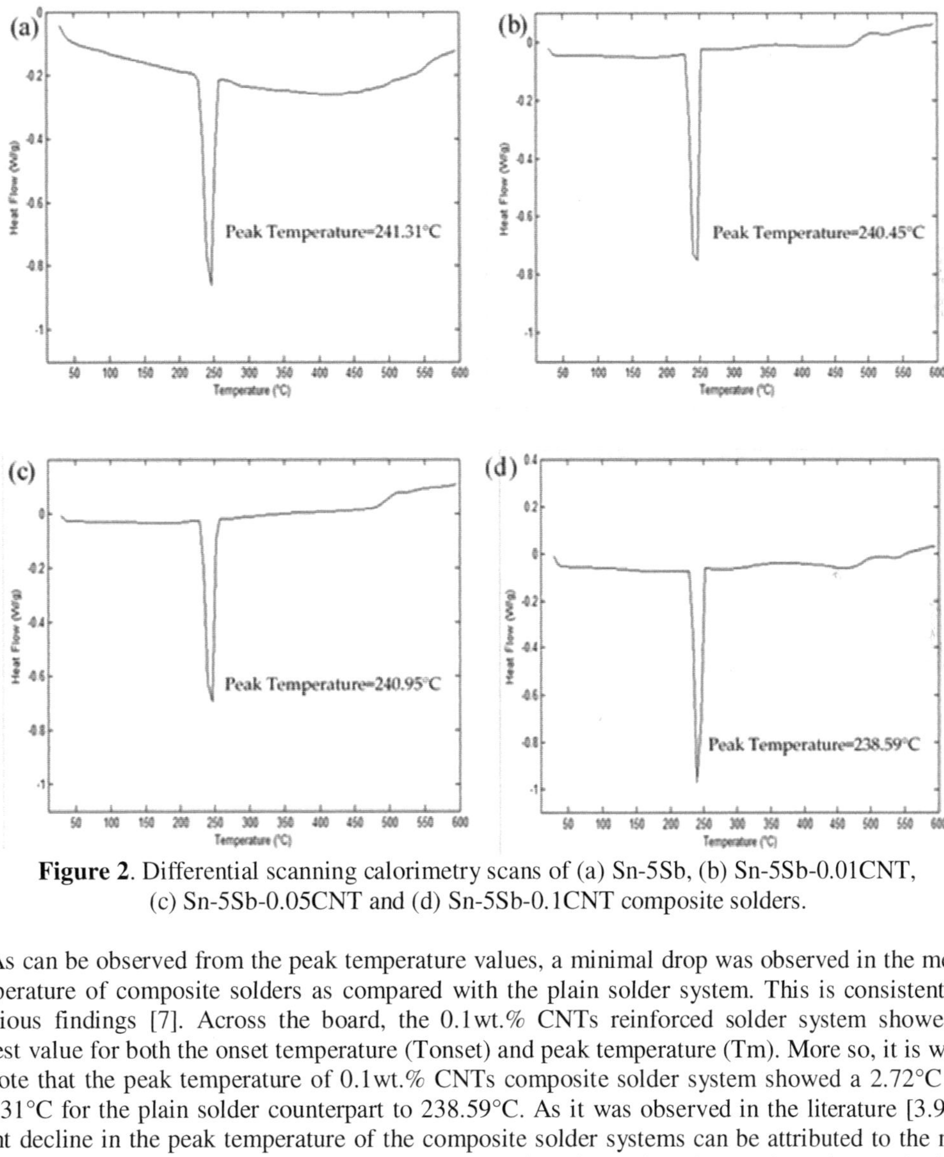

Figure 2. Differential scanning calorimetry scans of (a) Sn-5Sb, (b) Sn-5Sb-0.01CNT, (c) Sn-5Sb-0.05CNT and (d) Sn-5Sb-0.1CNT composite solders.

As can be observed from the peak temperature values, a minimal drop was observed in the melting temperature of composite solders as compared with the plain solder system. This is consistent with previous findings [7]. Across the board, the 0.1wt.% CNTs reinforced solder system showed the lowest value for both the onset temperature (Tonset) and peak temperature (Tm). More so, it is worthy of note that the peak temperature of 0.1wt.% CNTs composite solder system showed a 2.72°C from 241.31°C for the plain solder counterpart to 238.59°C. As it was observed in the literature [3.9], the slight decline in the peak temperature of the composite solder systems can be attributed to the multi-walled carbon nanotubes role in exciting surface instability in the host Sn matrix owing to the higher surface free energy potentials of the CNTs.

IOP Conf. Series: Materials Science and Engineering **238** (2017) 012010 doi:10.1088/1757-899X/238/1/012010

Sample	Peak Temperature T_m(°C)	Solidus Temperature T_{end} (°C)	Liquidus Temperature T_{onset}(°C)	Pasty Range
Sn-5Sb	241.31	253.42	231.80	21.62
Sn-5Sb-0.01CNT	240.45	253.00	231.36	20.64
Sn-5Sb-0.05CNT	240.95	253.39	231.98	21.41
Sn-5Sb-0.1CNT	238.59	250.33	230.97	19.36

Table1. Thermal parameters for plain and composite solders from DSC scans.

3.2. Interfacial IMC formation

The representative optical micrographs and the SEM-EDX evaluation for the intermetallic evolution results of as reflow and 500h aged samples are presented in Figures 3 and 4 respectively. From the optical micrographs of both the as-reflow and aging conditions, a remarkable IMC layer growth retardation can be observed in the composite solder system. In particular, the composite solder system having 0.05wt.% CNTs reinforcement showed superior IMC layer growth retardation for all subjected conditions as compared with other solder systems.

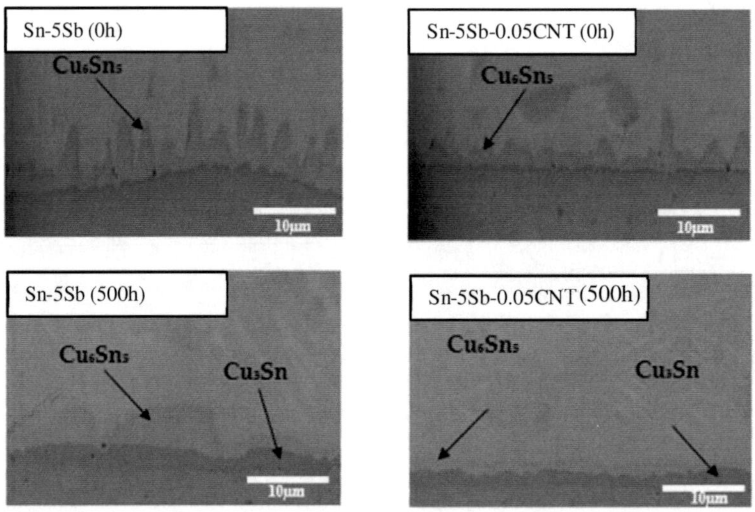

Figure 3. Optical micrographs of interfacial IMC layer of the as reflow and 500h aged solder joint samples.

Figure 5 presents the plot of total IMC layer thickness against different solder systems, wherein the degree of IMC layer growth retardation can be observed in the composite solder systems. For the Sn-5Sb plain solder and Sn-5Sb-0.05CNT composite solder, IMC thicknesses of 3.04μm and 2.60μm were recorded respectively for both solder formulations. In a similar way, the IMC layer thickness recorded for the Sn-5Sb and Sn-5Sb-0.05CNT after 500h of isothermal aging were 7.69μm and 6.73μm respectively which indicates a drop of 0.96μm in the composite sample as compared to that of the plain solder. Therefore, the IMC layer growth suppression experienced in the composite solder formulations can be attributed to the inhibitory role played by multi-walled carbon nanotubes in impeding Sn atoms diffusion from the solder matrix which is needed in the growth of the underlying Cu3Sn IMC layer. This finding is in line with the previous investigations [17,18].

IOP Conf. Series: Materials Science and Engineering **238** (2017) 012010 doi:10.1088/1757-899X/238/1/012010

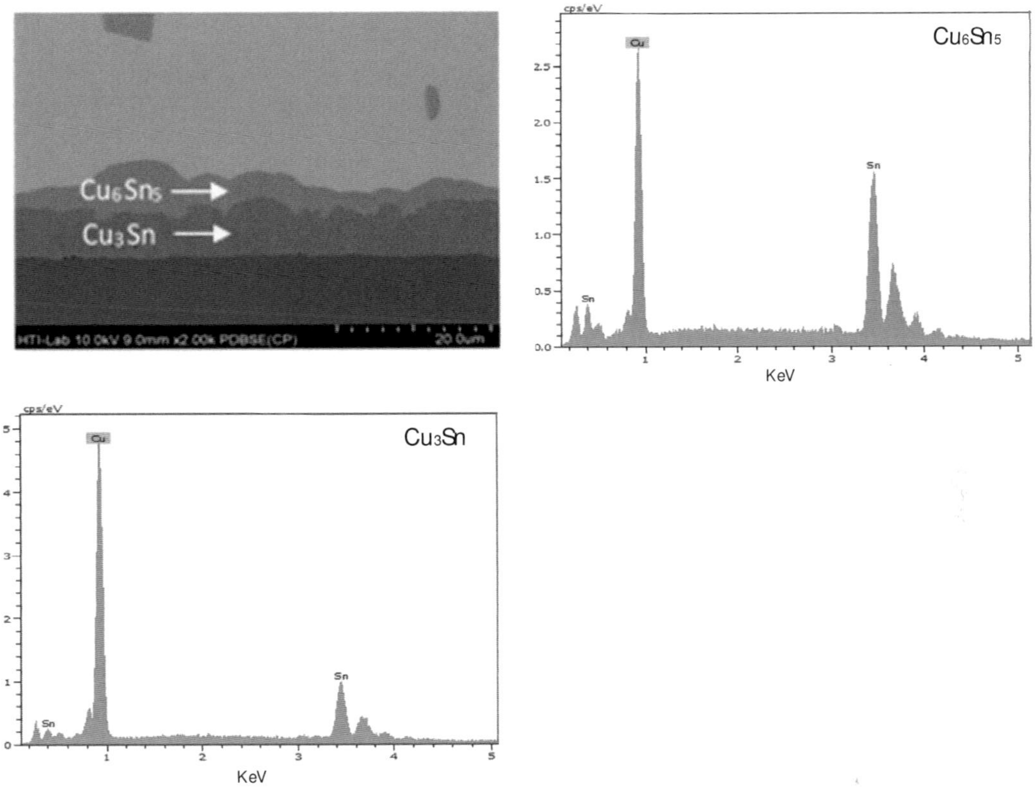

Figure 4. FESEM micrograph of the aged solder joint and the respective EDX spectrums of the interfacial IMC layers.

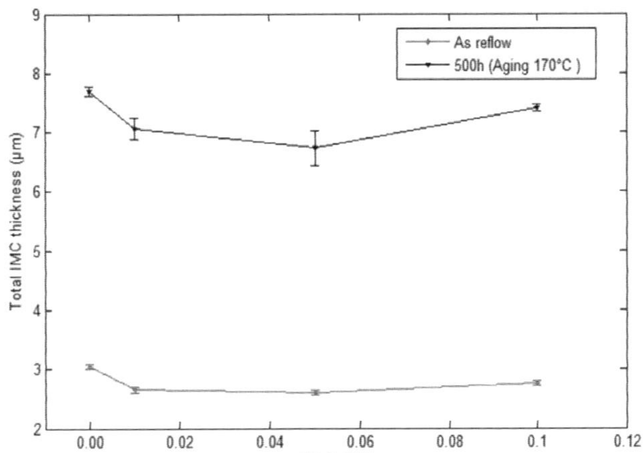

Figure 5. Overall IMC layer thickness against different solder systems subjected to as- reflow and aging conditions.

4. Conclusion

The Sn-5Sb solder systems with different weight fractions of multi-walled carbon nanotubes were fabricated by utilizing the powder metallurgy process. The melting properties and interfacial IMC layers for both the plain and composite solder systems were studied. The peak temperature of composite solder systems indicated a slight decrease which is indicative of MWCNTs role in exciting surface instability in the solder matrix. However the reduction was negligible and has no tendency towards disrupting the already established soldering routine of Sn-5Sb solder during electronic hardware packaging. More so, the interfacial IMC thickness of plain and composite solder joints were compared. It was observed that the as reflowed samples showed a marginal IMC layer suppression in the composite solder formulations as against that of the plain solder. However, the IMC layer growth suppression became remarkable in the composite solders after 500h of isothermal aging. This suggests that CNTs existence in solder matrix was effective in subduing the IMC layer growth.

References

[1] Dele-Afolabi T T, Hanim M A, Norkhairunnisa M, Yusoff H M and Suraya M T 2015. Investigating the effect of isothermal aging on the morphology and shear strength of Sn-5Sb solder reinforced with carbon nanotubes *J. Alloys Comp.* 649 368-374

[2] Nai S M L, Wei J and Gupta M 2006 Improving the performance of lead-free solder reinforced with multi-walled carbon nanotubes *Mater. Sci. Eng. A* 423 166–169

[3] Kumar K M, Kripesh V and Tay A A O 2008b Single-wall carbon nanotube (SWCNT) functionalized Sn–Ag–Cu lead-free composite solders *J. Alloys Comp.* 450 229–237

[4] Laurila T, Hurtig J, Vuorinen V and Kivilahti J K 2009 Effect of Ag, Fe, Au and Ni on the growth kinetics of Sn–Cu intermetallic compound layers *Microelectron. Reliab.* 49 242–247

[5] Shen J, and Chan Y C 2009 Microelectronics Reliability Research advances in nano-composite solders *Microelectron Reliability* 49 223–234

[6] Salleh M M, McDonald S D and Nogita K 2017 Effects of Ni and TiO 2 additions in as-reflowed and annealed Sn0. 7Cu solders on Cu substrates *J. Mater. Process Technol.* 242 235-245

[7] Kumar K M, Kripesh V, Shen L and Tay A A O 2006 Study on the microstructure and mechanical properties of a novel SWCNT-reinforced solder alloy for ultra-fine pitch applications *Thin Solid Films*, 504 371–378

[8] Kumar K M, Kripesh V and Tay A A O 2008a Influence of single-wall carbon nanotube addition on the microstructural and tensile properties of Sn–Pb solder alloy *J. Alloys Comp.* 455 148–158

[9] Niranjani V L, Rao B S S C, Singh V and Kamat S V 2011 Influence of temperature and strain rate on tensile properties of single walled carbon nanotubes reinforced Sn–Ag–Cu lead free solder alloy composites *Mater. Sci. Eng.: A* 529 257–264

[10] Han Y D, Nai S M L, Jing H Y, Xu L Y, Tan C M and Wei J 2010 Development of a Sn–Ag–Cu solder reinforced with Ni-coated carbon nanotubes *J. Mater. Sci.: Mater. Electron* 22 315–322

[11] Mayappan R, Hassan A A, Ghani N A, Yahya I and Andas J 2016 Improvement in intermetallic thickness and joint strength in carbon nanotube composite Sn-3.5 Ag Lead-free Solder *Mater. Today* 3 1338-1344

[12] Xu S, Cheong Y, Zhang K and Yung K C 2014 Interfacial intermetallic growth and mechanical properties of carbon nanotubes reinforced Sn3 . 5Ag0 . 5Cu solder joint under current stressing *J. Alloys Comp.* 595 92–102

[13] Che F X, Zhu W H, Poh E S, Zhang X W and Zhang X R 2010 The study of mechanical properties of Sn–Ag–Cu lead-free solders with different Ag contents and Ni doping under different strain rates and temperatures *J. Alloys Comp* 507 215-224

[14] Zeng G, Xue S, Zhang L, Gao L, Dai W and Luo J 2010 A review on the interfacial intermetallic compounds between Sn–Ag–Cu based solders and substrates *J. Mater. Sci.: Mater. Electron* 21 421-440

[15] Suganuma K, Kim S and Kim K 2009 High-Temperature lead-free solders : properties and possibilities *J. Mater. Sci.* 61 64-71
[16] Frear D R and Vianco P T 1994 Intermetallic growth and mechanical behavior of low and high melting temperature solder alloys Mettalurg. *Mater. Trans. A 25* pp 1509–1523
[17] Yang Z, Zhou W and Wu P 2014 Effects of Ni-coated carbon nanotubes addition on the microstructure and mechanical properties of Sn – Ag – Cu solder alloys *Mater. Sci. Eng. A* 590 295–300
[18] Nai S M L, Wei J and Gupta M 2009 Interfacial intermetallic growth and shear strength of lead-free composite solder joints *J. Alloys Comp.* 473 100-106

Optimization on laser soldering parameters onto lead-free solder joint

T J Nabila*, S R A Idris, M Ishak

Faculty of Mechanical Engineering, Universiti Malaysia Pahang, 26600 Pekan, Pahang, **Malaysia**

E-mail: nabilatj_92@yahoo.com

Abstract. This paper presents the results of optimization on laser soldering parameters onto lead free solder joint. The objective of this study is to determine the laser power needed and scanning time required to produce a high quality of solder joint. Laser soldering was selected due to its rapid and controllable selective process as well as suitable for temperature sensitive assemblies, which is very desirable nowadays as compared to time consuming process such as casting. A fibre laser with 200W was used in this study to form a joining between lead free solder wire and copper board. However, a continuous laser power was ranged between 48 to 60W to create joining. Before that, flux was also applied prior to laser soldering for uniform heat distribution throughout the solder volume. Lead free solder wire with the size of 1.0mm diameter was formed into several shapes including straight line-shape with 20mm in length, and spiral-shape with 6mm diameter, in order to capture and disperse the heat evenly. Results showed that laser power ranged from 48W to 60W with 3.5 to 4.5s duration was found to be suitable for lead free solder wire with spiral-shape condition. Besides, the wetting angle was also in optimal when the laser power was increased The optimized fibre laser parameters obtained in this study will be used for future reference in performing a laser soldering between solder alloy copper board.

1. Introduction

Due to the environmental and health concerns, Sn-Pb solders have been banned and the studies on lead-free solder have been done world widely in conjuction of supporting Waste Electrical and Electronic Equipment (WEEE) and Restriction of the Use of Hazardous Substances in Electrical and Electronic Equipment (RoHS Directive) [1-3]. Lead-free solder such as Sn-Cu solder alloys are seen as one of great potential substitute of Sn-Pb solder alloys due to it's good wettability, high electrical conductivity, good mechanical properties and economically costing [4-6]. Furthermore, Cu has high availabity, good electrical and thermal conductivity, therefore it has always been chosen for any metallization layers medium in the manufacturing of electrical components [7, 8].

Laser soldering technology is known to have been used since many years before due to it's highly advanced selective soldering whereas the heat is applied only on the desired spot, mechanically controlled by the machine and thus result in high production rate [2, 8, 9]. Meanwhile, as to compare with conventional method which is reflow soldering, the laser soldering has contactless temperature measurement during laser process and rapid rise and fall temperature cycles that will certainly minimize the thermal damage especially for heat-sensitive components [10, 11]. It is also been claim that soldering by using laser technology produce a good electrical conductivity in the integrated circuits assembly and void defects are minimized due to it's fast evaporation of flux [1, 12]. Therefore, laser solderimg technology has been quite well-known of it's advantages in electronic packaging industries however the study on it's optimal parameters are still ongoing.

IOP Conf. Series: Materials Science and Engineering **238** (2017) 012011 doi:10.1088/1757-899X/238/1/012011

However, finding the optimal laser parameters is difficult as the heat diffusion from solder to surrounding parts and beam reflection are unclear. For human work, they are able to control the heating by visual confirmation for precise soldering. However, laser soldering is completely mechanically controlled by the machine [8]. Meanwhile, if the rate of heating is too intensive and surpass the required temperature for the solder to melt, it may cause some deterioration to the materials surrounding, nearest components and layer surface and thus affects the electrical flow performance but when insufficient output power is applied, the solder itself may not be melt [2, 13]. Thus, it is essential to find the optimal laser parameters such as the laser output power and irradiation time in order to gain good joints and avoid unexpected complication later [13, 14].

Therefore, in this study, we performed some experiments about laser soldering in order to find the optimal parameters of lead free solder joint. In the experiment two types of shape which are straight line-shape and spiral-shape were used, in order to capture and disperse the heat evenly. Then, we discussed the solder shape dependence on heat absorption and the effects of different parameters on the wettability joint

2. Experimental Procedure

Lead free solder wire (diameter of 1.0 mm), SnCu was used in the experiment. The solder wire was cut in straight line shape with 20mm in length and was placed on the printed copper board (PCB) manually (**Figure 2** (a)). The PCB was clean first by using ultrasonic cleaner with ethanol as the cleansing solution. A thin layer of Kester TSF-6522RH rosin based flux was applied on the PCB. After the solder wire was placed on the PCB, a thin layer of flux was lightly applied again on top of the solder wire. The laser device used in this experient was a continuous wave type fiber laser (Figure 1). The laser parameters such as the output power and irradiation time were computer-controlled. The output power of the laser used are in range of between 48W to 60W and for the irradiation time it was varied to 3.5s, 4.0s and 4.5s. The PCB then are placed on the x-y platform jig as shown in Fig 1. The procedures are then repeated for spiral-shape with 6mm diameter solder wire (Figure 2 (b)).

After laser soldering, the samples were observed physically. The mechanical properties that were observed such as the joining between the solder and the PCB and the wetting angle performances.

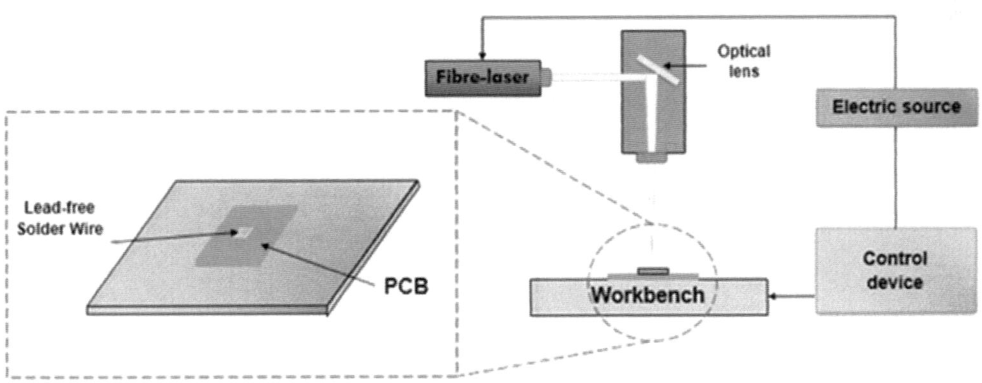

Figure 1. The illustration of experimental apparatus.

IOP Conf. Series: Materials Science and Engineering **238** (2017) 012011 doi:10.1088/1757-899X/238/1/012011

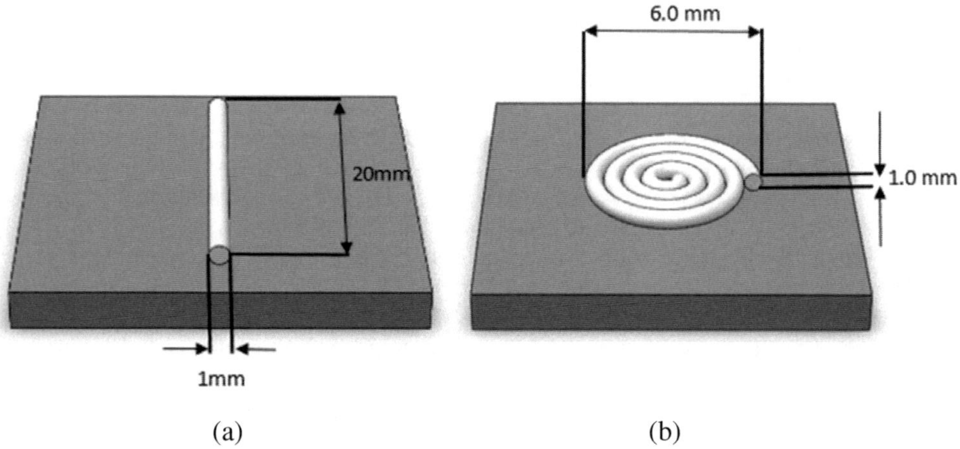

Figure 2. Illustration of the shape of solder wire on pad. (a) Straight-line and (b) spiral-shape with 6mm in diameter.

3. Results and Discussion

3.1. Effect of solderwire shape on the solder joint

Figure 3 and Figure 4 show the photos of top views of the solder wire that has been soldered by using the laser with different laser output power. When using a 50W laser power heating time for 3.5s, it was observed that the solder was not joint on the pad for a straight-line solder wire shaped (Figure 3 (a)). Meanwhile, when the power was increased to 55W and 60W, it could be seen that, the solder did created joint with the pad (Figure 3 (b) and (c)). However, the surface wetting for both joint were no good and the solders were almost burned. When the power is increased, the solder joint increased however, at a certain power output, the heat might be too much and caused burned on the solder.

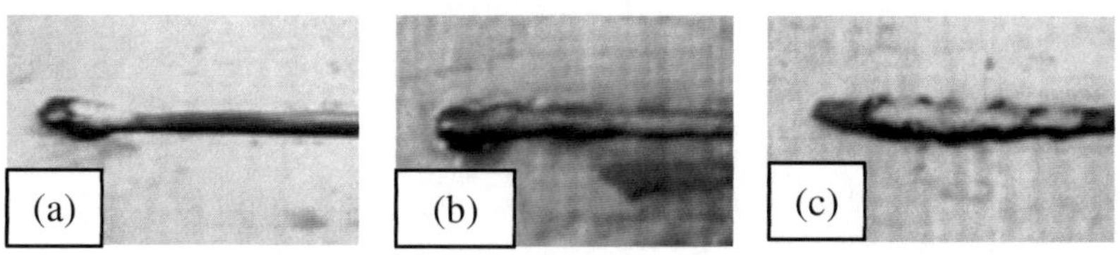

Figure 3. The top views of straight-line in shaped solder wire, laser heating time of 3.5s and laser output power of (a) 50W, (b) 55W and (c) 60W.

For the spiral in shaped solder wire, the solder did create a joining when the laser output was ranged in between 48W to 60W and irradiation time of 3.5s. As it can be seen, as the power output was increased (from left to right) the area of the solder that are soldered was increased. Good joining between the solder and the substrate interface could be gained, when larger contact area are created on them [15]. Furthermore, the spiral-shape solder has a larger volume of solder compared to a straight-line shape solder. The heat are able to disperse evenly before it solidifies completelety when volume is increased. Therefore, the larger contact area between solder/surface and volume of solder are some factors that affect a solder joint.

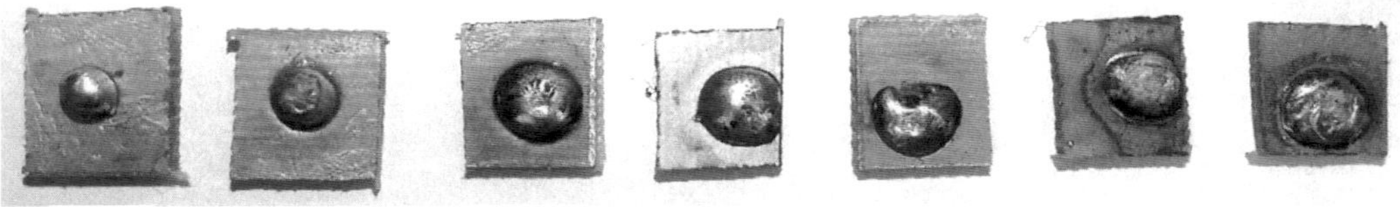

(a) (b) (c) (d) (e) (f) (g)

Figure 4. The top views of spiral in shaped solder wire, laser heating time of 3.5s and laser outputpower of (a) 48W, (b) 50W, (c) 52W, (d) 54W, (e) 56W, (f) 58W and (g) 60W.

3.2. Effect of different laser parameters on the wettability of solder joint

Wettability of solder joint plays an important role in the manufacturing of electrical components as it is the measurement ability for the molten solder to make interconnection between solder and substrate [16, 17]. The acceptable contact angle range should be below 45° and the smaller the contact angle the better is it's wettability joint [18]. In Figure 5 shows the wetting angle of spherical-shape lead-free solderwire with irradiation time of 4s and different laser power output range in between 48W to 60W. The wetting angle was decreases as the laser power output increasing from 48W to 52W. However, the wetting angle starts to increases from 54W to 60W of laser power output. It can be said, that the optimal wetting angle was found in between range of 52W to 60W power output.

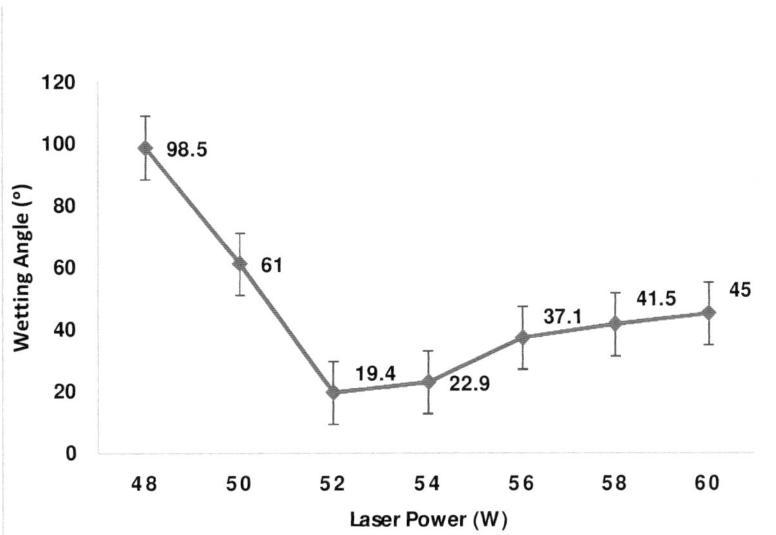

Figure 5. The relationship between wetting angle (°) and the laser power (W) for spherical-shape lead-free solder wire.

Figure 6 shows the wetting angle of spherical-shaped lead-free solder when the laser power output was fixed for 52W and the irradiation times were varied from 3.5s to 4.5s. When the irradiation time is 3.5s, the wetting angle, 44.1° was near to the optimal range of wetting angle as mentioned before. However, for 4s irradiation time, the solder molten can be seen to be more flatten compared to the others with wetting angle of 19.4°. At 4.5s irradiation time, the wetting angle increases again up to 37.6°, but still in optimal range as claimed from previous studies. For all three solders, it shows that the wetting angle gained are still in the range of good wetting angle which is below 45°.

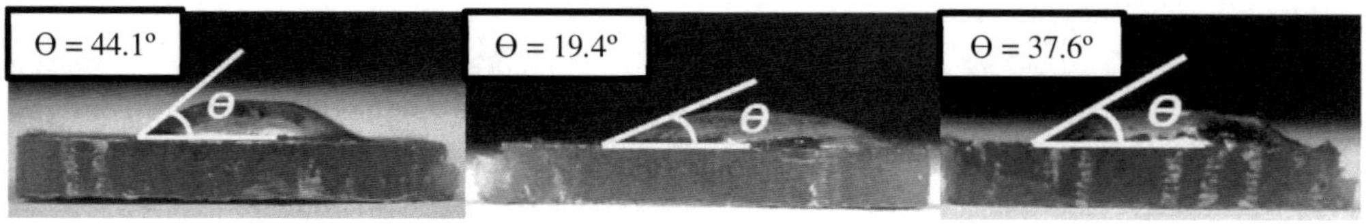

Figure 6. The wetting angle for spherical-shaped lead free solder wire with laser power output of 52W and irradiation time of (a) 3.5s, (b) 4s and (c) 4.5s.

It is shown here, the weting areas of solder on pads were getting larger as the output laser power is increase. As the output laser power increased, the area of solder to melt getting larger plus the contact areas between the solder and the surface of the substrate increased correspondingly making the solder spherical shape to disappear slowly and become more flatten, thus larger wetting areas are gained [19]. Meanwhile, when the irradiation time for solder is insufficient, the solder molten solidified faster and gives out not enough time for solder to disperse more on the surface of substrate thus making the wetting angle decreases. Therefore, in order to achieve an acceptable wetting, a balance between power output and heating time is inevitable.

4. Conclusion

In this study, it can be concluded that:

i. The contact area and the volume of the solder affects the rate of joining between the solder and substrate. The larger the contact area, the better the solder to joint. As in this study, the spherical-shape lead-free solder wire results in better joining compared to the straight-line shape lead-free solder wire.

ii. The increases of power output, increases the wettability joint of solder. However, a balance of controlling the power output and irradiation time parameters should be taken in account in order to obtain a good wettability joint.

Acknowledgement

The authors are thankful for the facilities provided by the Fakulti Kejuruteraan Mekanikal, Universiti Malaysia Pahang.

References

[1] Jeng-O K, Jae-Pil, Jae-Hoon L, Suh J, Hee-Shin K, 2009 Effects of laser parameters on the characteristics of a Sn-3.5 wt.%Ag solder joint *Metals and Materials International* 15 (1) p 119-123

[2] Illyefalvi-Vitez Z, Balogh B, Baranyay Z, Farmer G, Harvey T, Kirkpatrik D, Kotora G, Ruzsics N, 2007, Laser Soldering for Lead-free Assembly. ISSE 2007 - 30th International Spring Seminar on Electronics Technology 2007 *Conference Proceedings: Emerging Technologies for Electronics Packaging*, p 471-476

[3] Wang M, Wang J, and Ke W, 2017 Corrosion behavior of Sn-3.0Ag-0.5Cu lead-free solder joints *Microelectronics Reliability* 73 p 69-75

[4] Abd El-Rehim A F and Zahran H Y 2017, Investigation of microstructure and mechanical properties of Sn-xCu solder alloys *Journal of Alloys and Compounds* 695, p 3666-3673

[5] Zeng G, Xue S, Zhang L, Gao L, Lai Z, Luo J, 2010, Properties and microstructure of Sn–0.7Cu–0.05Ni solder bearing rare earth element *Pr. Journal of Materials Science: Materials in Electronics* 22(8) p 1101-1108

[6] Zeng G, Xue S, Zhang L, Gao L, 2011, Recent advances on Sn–Cu solders with alloying elements: review *Journal of Materials Science: Materials in Electronics*, 22(6), p 565-578

[7] Yao P, Li X, Liang X, Yu B, 2017, Investigation of soldering process and interfacial

microstructure evolution for the formation of full Cu_3Sn joints in electronic packaging *Materials Science in Semiconductor Processing* 58 p 39-50
[8] Kibushi R, Hatakyeama T, Imai D, Nakagawa S, Ishizuka M, 2013 Optimal Laser Condition for Laser Soldering in Cream and Ring Solder
[9] Nicolics J, 1992 Optimization of Process Parameters for Laser Soldering of Surface Mounted Devices 15(6) p 1155-1159
[10] Nishikawa H, Noriya I., Takemoto T, Effect of heating method on microstructure of Sn-3.0Ag-0.5Cu solder on Cu substrate, *Proc. of 18th European Microelectronics Packaging Conference*, p 221-224
[11] Nishikawa H and Iwata N, 2015 Formation and growth of intermetallic compound layers at the interface during laser soldering using Sn – Ag Cu solder on a Cu Pad *Journal of Materials Processing Technology* 215 p 6-11
[12] Bunea R, Svasta P, Zsolt I, 2011 Optimizing Laser Soldering of SMD Components : From Theory to Practice p 55-58
[13] Beckett P M, Fleming A R, Gilbert J M, Whitehead D G, 2002 The finite element modelling of laser soldering for electronic assemblies, *Transactions of the Institute of Measurement and Control*, 21, p 265-281
[14] Wang Y, Xiong Z, Zou X, Ding H, 2006 Experimental Optimization of Process Parameters for Diode Laser Soldering of BGA
[15] Shao-cheng L, Yu-ping H, Yee-wen Y, 2009 Interfacial Reactions and Mechanism Properties between SAC 405 and SACNG Lead-free Solders with Au/Ni (P)/Cu Substrates Reflowed by CO_2 Laser p 300-302
[16] Wang L, Yu D Q, Zhao J, Huang M L, 2002 Improvement of wettability and tensile property in Sn-Ag-RE lead-free solder alloy 56 p 1039-1042
[17] Murad N, Idris S R A, Ishak M, 2016, Effects of sintering temperatures on microstructures and mechanical properties of Sn4.0Ag0.5Cu1.0Ni solder alloy, *Electronics Manufacturing Technology (IEMT) & 18th Electronics Materials and Packaging (EMAP) Conference* p 1-7
[18] Mohd Salleh M M A, Mustafa Al Bakri A M, Kamarudin H, Bnhussain M, Zan Hazizi M H, Somidin F, 2011 Solderability of Sn-0.7Cu/SiN lead-free composite solder on Cu-substrate p 299-304
[19] Wei L, Chunqing W, Mingyu L, Li L, 2005, Effect of Laser Input Energy on Wetting Areas of Solder and Formation of Intermetallic Compounds at Sn-3.5Ag-0.75Cu/Au Right-angled Joint Interface 2 p 2-6

Effect of Zinc Additions on Sn-0.7Cu-0.05Ni Lead-Free Solder Alloy

N A Saleh, M I I Ramli and M A A Mohd Salleh*

Centre of Excellence Geopolymer and Green Technology, Scholl of Materials Engineering, Universiti Malaysia Perlis (UniMAP), Taman Muhibbah 02600, Jejawi, Arau, Perlis , Malaysia.

E-mail: arifanuar@unimap.edu.my

Abstract. This study focuses on the effect of zinc additions on Sn-0.7Cu-0.05Ni lead free solder alloy. Micro-alloying with additions of 0.5, 1.0 and 1.5 wt.% of zinc (Zn) was developed by using conventional casting method. Samples were analyzed using optical microscope (OM), scanning electron microscope (SEM), X-ray diffraction (XRD), differential scanning calorimetry (DSC) and vickers hardness testing. Analysis were carried out to investigate the microstructure, thickness of intermetallic compounds (IMCs), wettability, phase formation, thermal properties and hardness of Sn-0.7Cu-0.05Ni solder alloys with additions of different Zn percentage. From this study, it is found that different composition of Zn affected the physical and mechanical properties of Sn-0.7Cu-0.05Ni solder alloys. Results shows that increasing Zn could increase the wettability, reduce the melting point and slightly reduce the ultimate tensile strength of the solder alloy.

1. Introduction

In this digital-age era, solder plays an important role in electronic packaging industries. Most electronic devices uses solder as interconnects material connecting one component to another. As interconnects material, solder provides electrical and mechanical support to the electronic device. Solder alloy usually consist of two or more addition of microalloying where soldering process is a method used to join the component with a filler metal by melting two or more of metal below their melting point. Nowadays, lead free solders has been a major demand in the electronic industries due to the health and safety concern where lead solder has been banned.

Due to environmental and health concern, the consumption of lead is forbidden in electronic packaging industry [1, 2]. One of the most popular lead-free solder alloy used in wave soldering process is Sn-0.7Cu-0.05Ni. The excellent property of Sn-0.7Cu-0.05Ni solder alloy is it has good fluidity which may result in good uniformity to the solder thickness on copper substrate after wave soldering. It is also less erosive to printed circuit board (PCB) compared to Sn-Ag-Cu solder alloys [3]. However, Sn-0.7Cu-0.05Ni has limitation of high melting point and mechanical properties [4, 5].

Thus, a way in improving the alloy, additions of Zn will be carried out. It is predicted that additions of Zn in Sn-0.7Cu-0.05Ni will improve the microstructure, mechanical properties and wettability [6, 7]. Literatures indicates that additions of Zn may reduce the melting point of solders [6]. It is also good for improvement of wettability which enhance the ability of molten solder to spread over the substrate (copper) [6]. In addition, Zn is considered cheap compared to other microalloying elements.

Different compositions of Zn may also result in different effects to intermetallic compounds (IMCs) phases, thermal properties and mechanical properties. In this study, different wt. % of Zn additions to Sn-0.7Cu-0.05Ni will be investigated. The microstructure, thermal and mechanical properties of the Zn microalloyed solder will be discussed.

2. Experimental Procedure

Solder alloy of Sn-0.7Cu-0.05Ni was used as a base material. The solder alloy samples were fabricated with additions of different composition (wt. %) of Zn by using conventional casting technique. The solder alloys were mixed with three different compositions (0.5, 1.0 and 1.5 wt. % of Zn). Base solder alloys were firstly heated above their melting point and maintained for one hour. Then, the temperature was increase to superheat temperature at 600 °C for Zn additions. Every ten minutes of constant temperature for 30 minutes, the mixing material was stirred for homogeneity and solidified to room temperature.

For microstructural analysis, samples were divided to (i) bulk sample and (ii) sample soldered on substrate microstructure analysis. For bulk sample microstructure analysis, cross-sections were done after casting. While for sample soldered on substrate microstructure analysis, cross sections were done after solder alloy samples were solder reflowed on a Cu substrate in a F4N reflow oven. Some activated rosin flux was applied during reflow soldering to act as a cleaning agent to remove and prevent the formation of oxides. All samples were mounted with epoxy resin, grinded and polished for microstructure observations. The microstructure observations were carried out by optical microstructure (OM) and scanning electron microscope (SEM). In this microstructure analysis, intermetallic compound thickness and wettability of solder alloys (as shown in Figure 1) were measured.

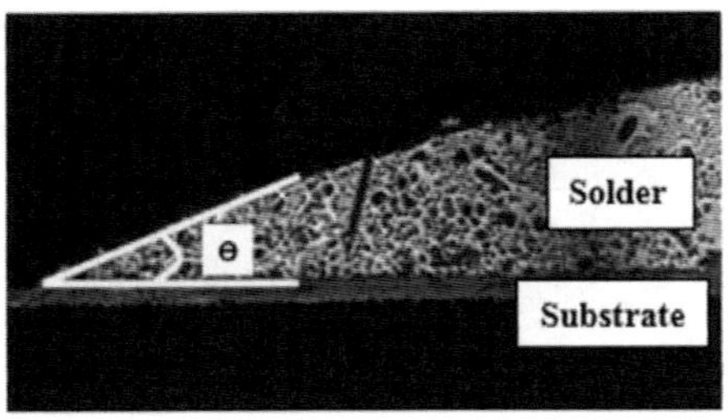

Figure 1. Wettability measurement by measuring the contact angle of the solder alloy soldered on Cu substrate.

For thermal analysis samples, differential scanning calorimeter (DSC) was used in investigating the thermal properties of the solder during soldering. Samples were in solder balls shape (600 μm diameter) and sample experiment setup were prepared as in figure 2. Samples were heated from room temperature to 250ºC at a heating and cooling rate of 10 ºC/min.

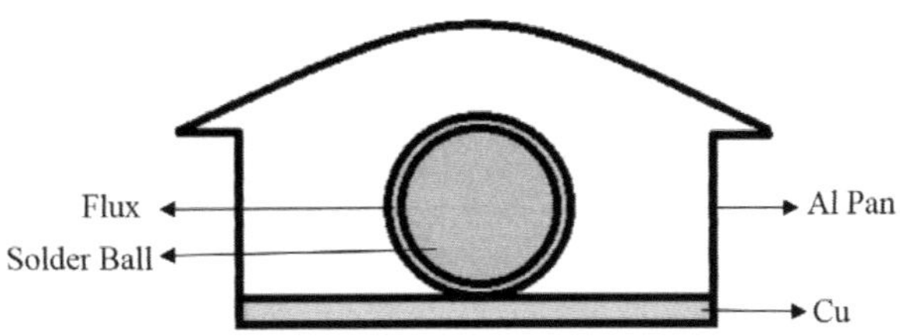

Figure 2. Sample preparation for differential scanning calorimeter (DSC) experiments.

For mechanical and phase analysis, analysis was conducted on bulk samples after casting. Vickers hardness indenter was used to measure the hardness of sample by using an indention load of 1kg.f at 10 s dwell time and 15 indentation for each sample was measured. For phase identification analysis, x-ray diffraction (XRD) was conducted with 0.1° step size at 25°- 80° range.

IOP Conf. Series: Materials Science and Engineering **238** (2017) 012012 doi:10.1088/1757-899X/238/1/012012

3. Results and Discussion

3.1. Bulk Microstructural Analysis

Figure 3 shows the optical micrograph of Sn-0.7Cu-0.05Ni solder alloy that containing 0, 0.5, 1.0 and 1.5 wt. % addition of Zn. One of the important solder material properties depends on their microstructure and based on studied by G. Zeng et al. [8], grain structure of the solidified alloy is depending on the solder composition.

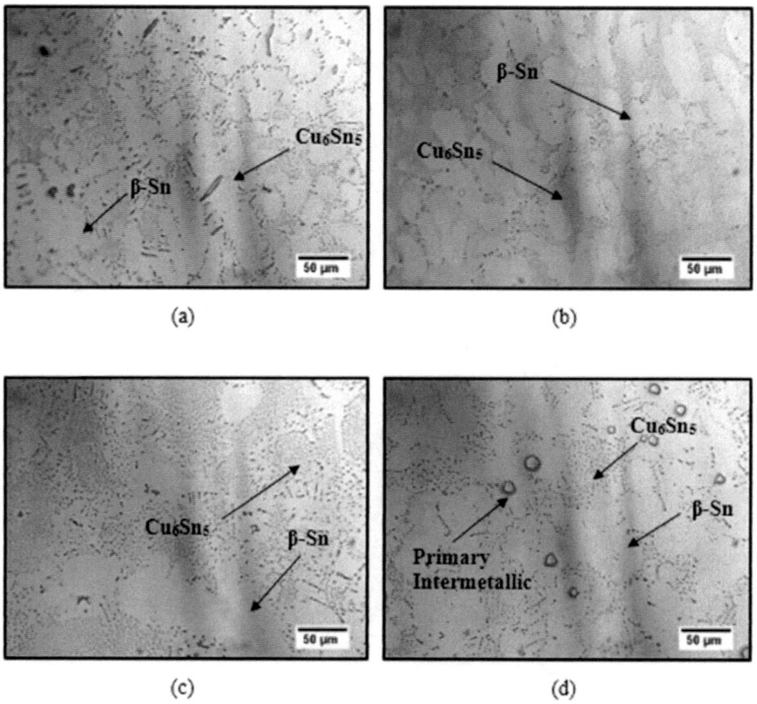

Figure 3. Optical micrograph of Sn-0.7Cu-0.05Ni solder alloy (a) Pure, (b) containing 0.5 wt.% Zn, (c) containing 1.0 wt.% Zn and (d) containing 1.5 wt.% Zn.

The typical microstructure Sn-0.7Cu-0.05Ni solder alloy that forms usually consist of primary β-Sn dendrites that surrounded by eutectic which is combination of β-Sn and intermetallic Cu_6Sn_5 [9]. W.Ng et al. [10] stated that Zn addition by itself did not make to a eutectic microstructure while Ni addition make the way it closer to eutectic microstructure. With the combination of Ni and Zn, its can refinement to near completely eutectic microstructure of the solder alloys [10-12]. As seen in Figure 3 (a), the pure Sn-0.7Cu-0.05Ni solder alloy consists a large number of primary β-Sn cells with small intermetallic compound Cu_6Sn_5. It also has uniform eutectic structure phase microstructure. On the other hand with 0.5 wt.% addition of Zn as shown in Figure 3 (b), the microstructure of β-Sn was refined and has smaller number of intermetallic compound Cu_6Sn_5

Interestingly, with a small addition of the Zn make the β-Sn cell become more refined compared to pure Sn-0.7Cu-0.05Ni solder alloy. The crystals forms can be considered to improve the reliability of solder joint. By the addition of Zn into the solder alloys, the primary β-Sn dendrites are still present in the microstructure but the eutectic been modified with the incorporation of Zn into the Cu_6Sn_5 crystal [10]. With the addition of 1.0 wt.% Zn into the Sn-0.7Cu-0.05Ni solder alloy as shown in Figure 3 (c), the microstructure become non-uniform because consist of very large β-Sn cells with larger number of intermetallic compound Cu_6Sn_5. The microstructure phase by addition of 1.5 wt. % of Zn also make a large β-Sn cell and smaller number of intermetallic compound Cu_6Sn_5 with primary intermetallic. This is because of the higher amount of Zn involves did not completely homogeneously dissolved in Sn-0.7Cu-0.05Ni solder alloy.

3.2. Interfacial Intermetallic Compounds (IMCs)

Figure 4 (a) show the average thickness of (Cu,Ni)6Sn5 intermetallic compounds (IMCs) are 10.7 µm higher than others. The addition of 0.5 wt% Zn in the solder system make the average thickness become 7.6 µm where was inadequate to suppress the intermetallic compounds (IMCs) formation. However, when 1.0 and 1.5 wt% Zn were added into the solder alloys, the graph show an increasing to IMC layer average thickness of the (Cu,Ni)6Sn5 intermetallic growth. Both of the composition only have smaller different average thickness value of layer which is 8.6 µm and 8.9 µm respectively. This is shows that with only small addition of the Zn was effective to suppress the IMCs thickness of the solder alloys.

In previous reviewed, G.Zeng et al. [8] stated that the performance of the interfacial IMCs layers can be improve by added minor alloying into solder. Based on I.Yahya et al. [13], Cu6Sn5 intermetallic is the first phase that formed on the copper substrate for all solder system with scallop structure. There has another formation of thin IMC layer that located between Cu6Sn5 intermetallic compounds (IMCs) and copper substrate. The Cu6Sn5 in the intermetallic is an unstable layer then it make Cu3Sn intermetallic compounds (IMCs) to form [13]. Figure 4 shows SEM images for as-soldered solders. Figure 4 (a) show the pure Sn-0.7Cu-0.05Ni solder alloy only contain (Cu,Ni)6Sn5 intermetallic compounds without any addition of Zn and have a balance scallop structure than others. In J. Yoon et al. [14] investigation, the form of $(Cu,Ni)_6Sn_5$ layer are form with of 4.3 at.% Ni [14].

Q.S.Zhu et al [15] observed that small addition of Zn in lead-free solders can depress the growth of intermetallic compounds (IMCs) which make the growth of Cu_3Sn and the formation of Kirkendall voids can be restrain [13, 15]. For figure 4 (b), (c) and (d), the un-balance scallop structure forms and the addition of Zn make the Cu3Sn become thinner. However, the different composition of Zn in the Sn-0.7Cu-0.05Ni solder alloy make (Cu,Ni)6Sn5 intermetallic compounds (IMCs) layer have different thickness. It has been confirmed by H.Wang et al. [16] that Zn would alloy interfacial Cu6Sn5 during interfacial reaction between the solder and substrate. In addition, the stabilizing effect by addition of Ni and Zn also can minimize the thermal expansion mismatch between the interfacial Cu_6Sn_5 and substrate [16].

IOP Conf. Series: Materials Science and Engineering **238** (2017) 012012 doi:10.1088/1757-899X/238/1/012012

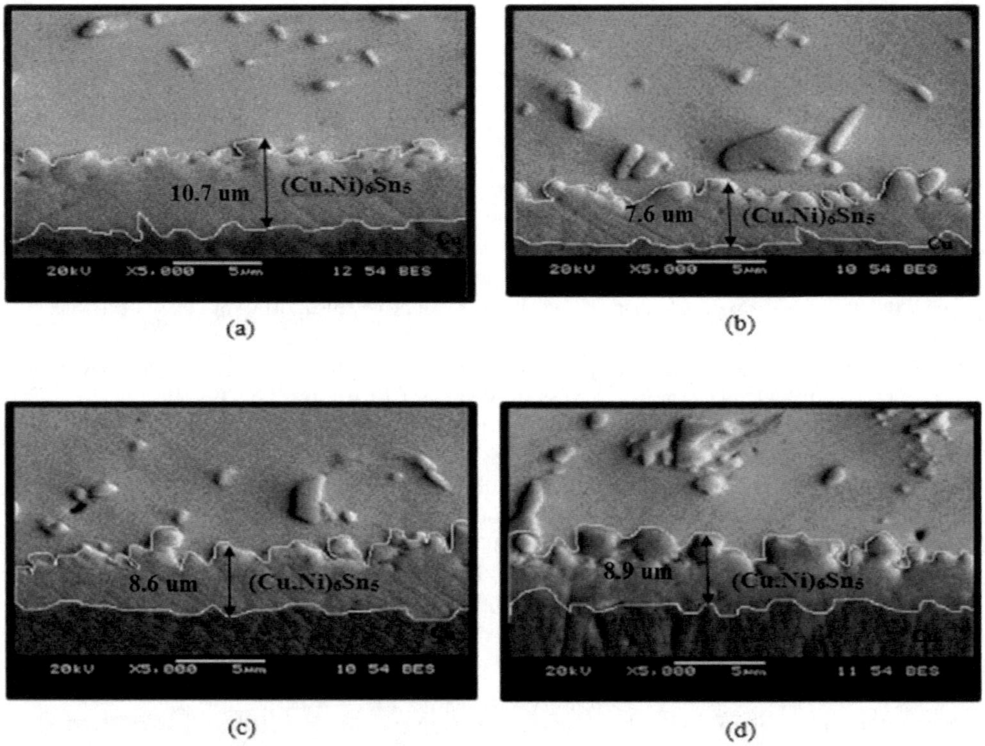

Figure 4. SEM images of Sn-0.7Cu-0.05Ni solder alloy (a) Pure, (b) containing 0.5 wt.% Zn, (c) containing 1.0 wt.% Zn and (d) containing 1.5 wt.% Zn.

3.3. Wettability

Wettability of solder alloys are depends on surface properties and also solder alloys proportion [17]. By minimize the value of the contact angle, the wettability can be optimized that corresponds to lower surface-interfacial energy. Based on the bar graph in figure 5, the contact angle for pure Sn-0.7Cu-0.05Ni solder alloy sample is 23.61° which is higher than 0.5 wt.% Zn but lower than 1.0 and 1.5 wt% addition of Zn. Y.Wang et al [18] explain that the addition of Ni into a solder alloys can improve the reliability and wetting properties of many Sn based solder alloys are well known [18, 19]. The addition of 0.5 wt% Zn in the solder system make the contact angle become 17.32° where much lower than others composition. J.Shen et al. [20] discovered that the addition of 0.01 and 0.03 wt.% Ni can decrease the wetting force and the wetting time of Sn-Cu solder cannot be effected [20]. This is show that the Ni and Zn have exactly same properties to lowered the wetting angle by addition of small amount of element.

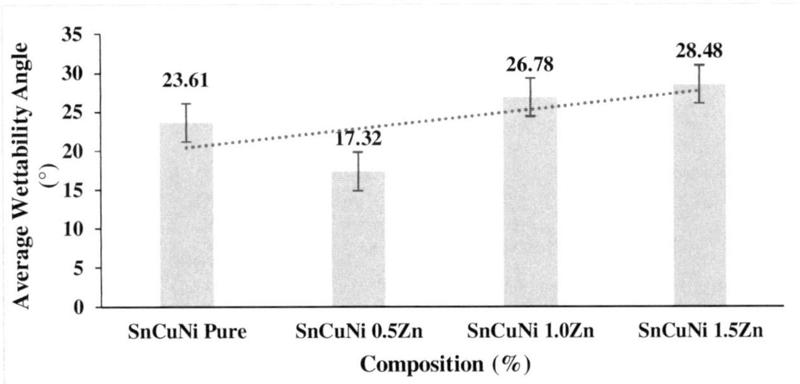

Figure 5. Bar graph of average wettability angle of Sn-0.7Cu-0.05Ni solder alloy with addition of 0.5, 1.0 and 1.5 wt.% Zn.

However, when 1.0 and 1.5 wt. % Zn was added into the solder alloys, the contact angle slightly increase into 26.78° and 28.48° respectively. L.Zang et al. [21] shared that improvement of the wettability on the Cu substrate with molten solder alloy are related with the formation of intermetallic [21]. By the formation of intermetallic that have been investigate for the addition of 1.0 and 1.5 wt.% of Zn, the higher thickness have been occurred and formed a higher wetting angle. In addition, alloying of Zn element into the Sn-based solder is concerned because Zn is very active and may deteriorate wetting performance of lead-free solder at molten state [16].

3.4. Phases Analysis

Figure 6 revealed the additional of another element in Sn-0.7Cu-0.05Ni solder alloys. For figure 6(a), only β-Sn and Cu6Sn5 phases are detected by XRD and there are no new addition of another element in the solder system. The crystal system occurred in this Sn-07Cu-0.05Ni solder alloy are tetragonal (Sn) and hexagonal (Cu6Sn5).

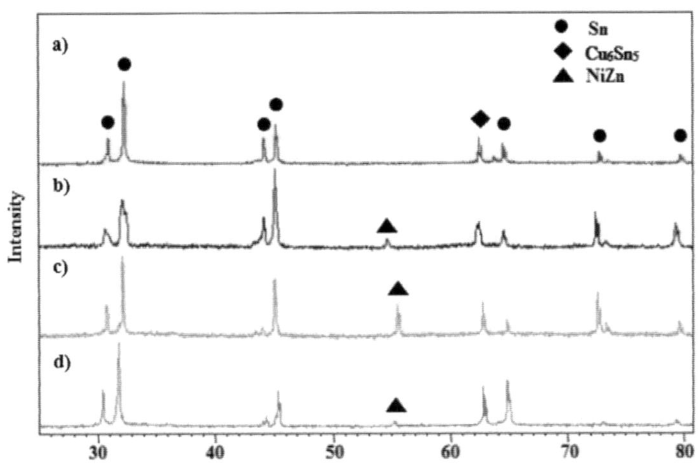

Figure 6. Peak of Sn-0.7Cu-0.05Ni solder alloys (a) Pure, (b) containing 0.5 wt.% Zn, (c) containing 1.0 wt.% Zn and (d) containing 1.5 wt.% Zn.

Figure 6 (b), (c) and (d) show that there is new peak are involved which is known as NiZn when different amount of Zn is added. The NiZn presence in the Sn-0.7Cu-0.05Ni solder alloy in cubic shape. According to G.Zeng et al.[11], the addition of Ni and Zn can effect over the polymorphic phase transformation of powdered Cu_6Sn_5 [11]. However, with the addition of different amount of Zn, the intensity β-Sn and Cu_6Sn_5 phase change are not fixed. It also believed that the addition of Zn can change the phase in solder system.

3.5. Phases Analysis

Melting temperature is a critical characterization for solder alloys because it determines the operating system temperature. Figure 7 shows the melting point of the Sn-0.7Cu-0.05Ni solder alloy with the addition of 0.5, 1.0 and 1.5 watts. % Zn. The Sn-0.7Cu-0.05Ni solder alloy with the addition of 0.5 wt.% have highest melting temperature with 230.93°C while the addition of 1.0 wt.% Zn have the lowest melting temperature which is 227.63 °C. This process is known as endothermic process because the process towards to heating process. T.T.Bao et al. [22] found interesting behavior in the phase diagram by using simple regular model which is claims when the positive interaction of solid and negative liquid occur, all the solidus and liquidous with the eutectic point move at the lower melting point component [22]. In addition, Y.Liu et al. [23] explain that the higher addition of Bi in the solder alloys could increase the melting range and can lead to the initiation of solidification crack of the solder joints [23].

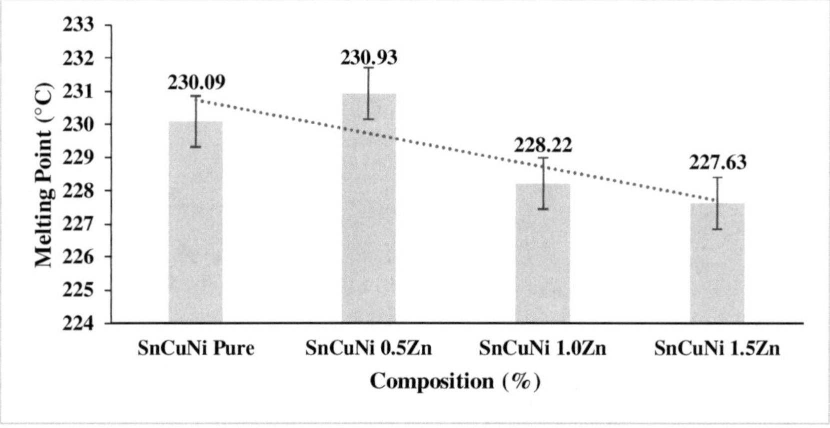

Figure 7. Melting point of Sn-0.7Cu-0.05Ni solder alloy with addition of 0.5, 1.0 and 1.5 wt. % Zn.

DSC result obtained the undercooling temperature has shown in the figure 8. Undercooling range temperature refer to solder alloys transfer from liquid to solid. The undercooling value can be calculated by minus the onset heating point with onset cooling point. Bar graph shows the higher temperature is on pure Sn-0.7Cu-0.05Ni solder alloy with 63.84°C compare to addition of 0.5 wt. % of Zn which have very low undercooling temperature. This is show that the optimum value for undercooling temperature is 0.5 wt. % addition of Zn because have low temperature different to change from liquid to solid.

For thermal analysis, it can be concluded that the low melting point achieved when 1.5 wt. % Zn are added. However, there are no slightly different value of melting point temperature for all samples. The melting point temperature are still around with eutectic or near-eutectic composition. For undercooling temperature, it shows that the addition of 0.5 wt. % of Zn has a short time to transfer from liquid to solid.

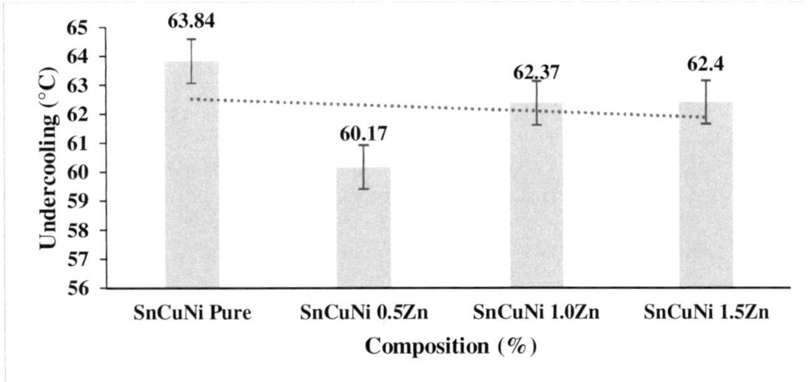

Figure 8. Undercooling of Sn-0.7Cu-0.05Ni solder alloy with addition of 0.5, 1.0 and 1.5 wt.% Zn.

3.6. Mechanical Properties

Figure 9 show the average of vickers hardness value for pure Sn-0.7Cu-0.05Ni solder alloy sample are 9.82 HV which is lower than addition of 0.5, 1.0 and 1.5 wt. % Zn. The addition of 0.5 wt. % Zn in the solder system make the higher value of hardness become 12.38 Hv. However, when 1.0 and 1.5 wt. % Zn was added into the solder alloys, the hardness slightly decreasing from 10.48 Hv to 9.87 Hv respectively. In fact, the motion of dislocation and growth and configuration of grains effected the micro-hardness of a solder alloys [24]. H.Wang et al. [17] concluded that the addition of Zn would induce the formation of CuZn and Cu6Sn5 which resulting to prevent plastic deformation and increase the hardness of Cu6Sn5 [17].

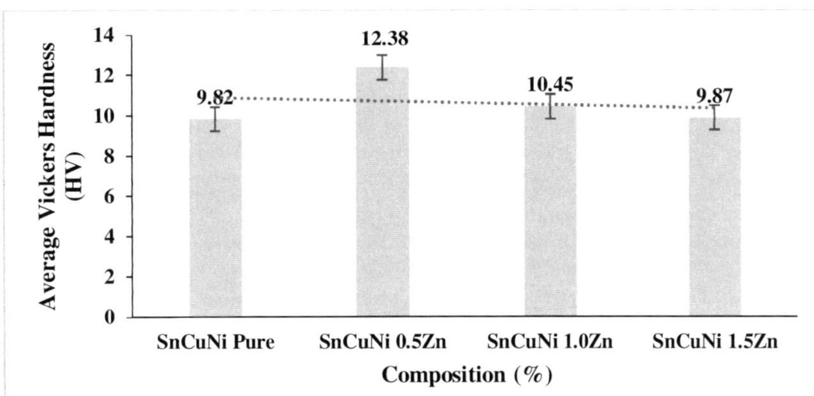

Figure 9. Average vickers hardness of Sn-0.7Cu-0.05Ni solder alloy with addition of 0.5, 1.0 and 1.5 wt. % Zn.

Figure 10 show the average ultimate tensile strength for pure Sn-0.7Cu-0.05Ni sample which is lower than addition of 0.5, 1.0 and 1.5 wt. % Zn. The addition of 0.5 wt. % Zn in the solder system make the higher value which is become 97.53 MPa. However, when 1.0 and 1.5 wt. % Zn was added into the solder alloys, the value of ultimate tensile strength slightly decrease into 92.58 MPa and 91.6 MPa respectively. The micro-hardness results of all the solder alloys system showed an improvement for certain amount of Zn.

The microstructural observation reveals that the distribution of Zn element through the refining of microstructure for each sample which can be considered as the main reason for the improvement in micro-hardness values of Sn-0.7Cu-0.05Ni solder alloy. It was concluded that Sn-0.7Cu-0.05Ni solder alloy with 0.5 wt.% Zn had the highest value of mechanical value among the four samples. It has also been observed that the mechanical value decreases with the increasing in the percentage of Zn content. However, when 1.0 and 1.5 wt% Zn was added into the solder alloy, the value of hardness slightly increase. Both of the composition only have small different value in the mechanical value. Mohd Salleh et al. [25] reported that the mechanical properties are affected by the grain size of the solder which the finer grains of microstructure have higher mechanical values [25]. In other studies, W.Ng et al. [10] claims that the addition of Zn can increase the mechanical properties of solder but when the Zn addition is combined with Ni, it can refine the eutectic microstructure which give beneficial effect [10].

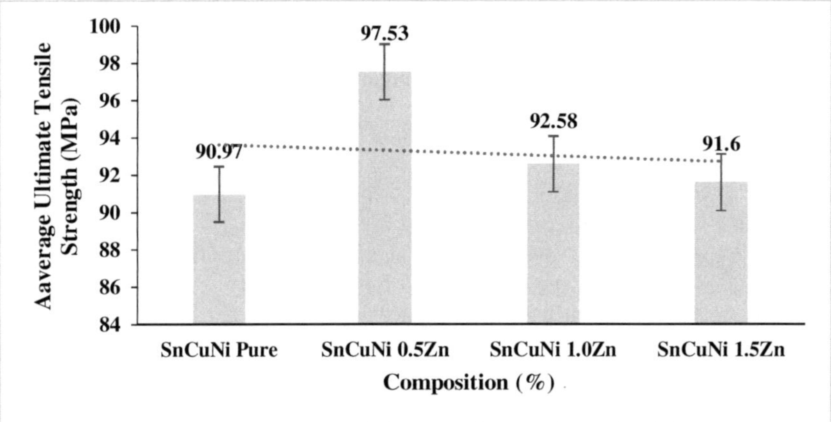

Figure 10. Average ultimate tensile strength of Sn-0.7Cu-0.05Ni solder alloy with addition of 0.5, 1.0 and 1.5 wt. % Zn.

4. Conclusion

The effect of Zn additions on Sn-0.7Cu-0.05Ni lead free solder alloy was examined through this study. The solder alloys were mixed with three different compositions (0.5, 1.0 and 1.5 wt.% of Zn) by using casting technique. The conclusions which can be drawn from this study are:

i. Large primary β-Sn dendrites with smaller number of the primary Cu_6Sn_5 were observed. Therefore, by additions of Zn, Cu_6Sn_5 could be refined and become smaller.
ii. By additions of Zn in Sn-0.7Cu-0.05Ni solder alloy, the thickness of interfacial $(Cu, Ni)_6Sn_5$ can be reduce.
iii. Mechanical properties increases with small wt.% of Zn content in Sn-0.7Cu-0.05Ni solder alloy.
iv. For thermal analysis, there are changes in the melting point and undercooling temperature of solder alloys after addition of 1.5 wt. % Zn.
v. Overall, Sn-0.7Cu-0.05Ni solder alloy with addition of 0.5 wt.% Zn were observed to be the most preferred composition compared to other Zn percentage.

References

[1] Somidin F, Mohd Salleh M A A, and Ahmad K R 2013 *Adv. Mater. Res* 620 105-111
[2] Mohd Salleh M A A, Hazizi M H, Mohd Abdullah M A B, Noriman N Z, Mayapan R, Ahmad Z A 2012 *Adv. Mater. Res* 626 791-796
[3] Faure C and Bath J, Lead-Free Wave Soldering, in Lead-Free Soldering, Bath J., Editor. 2007, Springer US. pp 91-116
[4] Ramli M I I, Saud N, Mohd Salleh M A A, Derman M N, Mohd Said R, Nasir N, *Mater. Sci Forum* 2015 161-166
[5] Mohd Salleh M A A, McDonald S, and Nogita K 2013 *App. Mech. Mater* 421 260-266
[6] Sun L and Zhang L 2015 *Adv. Mater Sci.Eng*, 2015
[7] Puttlitz K J and Stalter K A, Handbook of lead-free solder technology for microelectronic assemblies. 2004 CRC Press

[8] Zeng G, Xue S, Zhang L, Gao L, Dai W, and Luo J 2010 *J. Mater. Sci.Mater. Elec.* 21(5) 421-440
[9] Felberbaum M, Ventura T, Rappaz M, and Dahle A K 2011 *JOM*, 63(10) 52-55
[10] Ng W, Zeng G, Nishimura T, Sweatman K, McDonald S D, and Nogita K 2015 Electronic Packaging and iMAPS All Asia Conference (ICEP-IACC) International Conference on IEEE p. 809-813
[11] Zeng G, McDonald S D, Gu Q, Terada Y, Uesugi K, Yasuda H, and Nogita K 2015 *Act. Mater*, 83(0) 357-371
[12] Mohd Salleh M A A, McDonald S D, Gourlay C M, Yasuda H, and Nogita K 2016 *Mater. Des.* 108 418-428
[13] Yahya I, Ghani N A A, Mohd Salleh M A A, Abd Hamid H, Ahmad Z A, and Mayappan R 2012 *Electronic Manufacturing Technology Symposium (IEMT) 35th IEEE/CPMT International IEEE.* 1-6
[14] Yoon J-W, Lee Y-H, Kim D-G, Kang H-B, Suh S-J, Yang C-W, Lee C-B, Jung J, Yoo C-S, and Jung S-B 2004 *J. Alloy Compd*, 381(1) 151-157
[15] Zhu Q S, Song H Y, Liu H Y, Wang Z G, and Shang J K 2009 Electronic Packaging Technology & High Density Packaging ICEPT-HDP'09. *International Conference on IEEE* 1043-1046
[16] Wang H, Fang J, Xu Z, and Zhang X 2015 *J. Mater. Sci.Mater. Elec.* 26(6) 3589-3595
[17] Wang L, Yu D Q, Zhao J, and Huang M L 2002 *Mater. Lett.* 56(6) 1039-1042
[18] Wang Y, Wang G, Song K, and Zhang K 2017 *Mater. Des.* 119 219-224
[19] Sona M and Prabhu K N 2013 *J. Mater. Sci.Mater. Elec* 24(9) 3149-3169
[20] Shen J, Pu Y, Wu D, Tang Q, and Zhao M 2015 *J. Mater. Sci.Mater. Elec* 26(3) 1572-1580
[21] Zang L, Yuan Z, Zhao H, and Zhang X 2009 *Mater Lett.* 63(23) 2067-2069
[22] Bao T T, Kim Y, Lee J, and Lee J G 2010 Mater Trans. 51(12) 2145-2149
[23] Liu Y, Sun F L, Yan T L, and Hu W G 2008 *Electronic Packaging Technology & High Density Packaging, 2008. ICEPT-HDP International Conference on IEEE* 1-4.
[24] Islam R A, Wu B Y, Alam M O, Chan Y C, and Jillek W 2005 *J. Alloys Compd* 392(1) 149-158
[25] Salleh M A A M, Al Bakri A M M, Somidin F, Sandu A V, Saud N, Kamaruddin H, McDonald S D, and Nogita K 2013 Rev. Chim 64(7). 725-728.

IOP Conf. Series: Materials Science and Engineering **238** (2017) 012013 doi:10.1088/1757-899X/238/1/012013

A Study on The Development of Local Exhaust Ventilation System (LEV's) for Installation of Laser Cutting Machine

S. I. Harun, S. R. A. Idris*, N. Tamar Jaya

Faculty of Mechanical Engineering, Universiti Malaysia Pahang, 26600 Pekan, Pahang, Malaysia.

E-mail: rabiatull@ump.edu.my

Abstract. Local exhaust ventilation (LEV) is an engineering system frequently used in the workplace to protect operators from hazardous substances. The objective of this project is design and fabricate the ventilation system as installation for chamber room of laser cutting machine and to stimulate the air flow inside chamber room of laser cutting machine with the ventilation system that designed. LEV's fabricated with rated voltage D.C 10.8V and 1.5 ampere. Its capacity 600 ml, continuously use limit approximately 12-15 minute, overall length LEV's fabricated is 966 mm with net weight 0.88 kg and maximum airflow is 1.3 meter cubic per minute. Stimulate the air flow inside chamber room of laser cutting machine with the ventilation system that designed and fabricated overall result get 2 main gas vapor which air and carbon dioxide. For air gas which experimented by using anemometer, general duct velocity that produce is same with other gas produce, carbon dioxide which 5 m/s until 10 m/s. Overall result for 5 m/s and 10 m/s as minimum and maximum duct velocity produce for both air and carbon dioxide. The air gas flow velocity that captured by LEV's fabricated, 3.998 m/s average velocity captured from 5 m/s duct velocity which it efficiency of 79.960% and 7.667 m/s average velocity captured from 10 m/s duct velocity with efficiency of 76.665%. For carbon dioxide gas flow velocity that captured by LEV's fabricated, 3.674 m/s average velocity captured from 5 m/s duct velocity which it efficiency of 73.480% and 8.255 m/s average velocity captured from 10 m/s duct velocity with efficiency of 82.545%.

1. Introduction

Local exhaust ventilation (LEV) is an engineering system frequently used in the workplace to protect operators from hazardous substances [1]. To have an effective system it is important that it is well designed and installed, used correctly and properly maintained. Study this case and work on it with design and development to provide an effective system. Common problems nowadays are supervisor are often unaware their operators are being over-exposed to hazardous substances or that existing controls may be inadequate [2]. Besides that, sources of exposure are missed and the supervisor (and suppliers) are over-optimistic about the effectiveness of the controls. Moreover, existing controls have deteriorated also controls are not used correctly [3]. Need to work and study on designing and developing LEV system to avoid expensive mistakes and to control exposure effectively. Must provide LEV that is fit for purpose, is shown to work and continues to work. Ensure controls are adequate. Together with help of supervisor discuss on project must be competent in the operation of the LEV system [4].

Laser cutting is a technology that uses a laser to cut materials, and is typically used for industrial manufacturing applications, but is also starting to be used by schools, small businesses, and hobbyists [5]. Generation of the laser beam involves stimulating a lasing material by electrical discharges or lamps within a closed container [6]. As the lasing material is stimulated, the beam is reflected internally by means of a partial mirror, until it achieves sufficient energy to escape as a stream of monochromatic coherent light. Mirrors or fiber optics are typically used to direct the coherent light to a lens, which focuses the light at the work zone [7]. Thermal effects are the predominant cause of laser radiation injury, but photo-chemical effects can also be of concern for specific wavelengths of laser radiation. Even moderately powered lasers can cause injury to the eye. High power lasers can also burn the skin. Some lasers are so powerful that even the diffuse reflection from a surface can be hazardous to the eye [8]. Argon-ion lasers emit light in the range 351–528.7 nm. Depending on the optics and the laser tube a different number of lines is usable but the most commonly used lines are 458 nm, 488 nm and 514.5 nm. Laser line filters are narrow bandpass filters centered on the resonance of the laser, which attenuate the background plasma and secondary emissions that often result in erroneous signals [9].

2. Experimental Procedure

2.1. Design the selected LEV's

A specification should be drawn up for the operator's review. The specification should consider the work processes, the operator involved, the potential fugitive dusts/fumes/gases and the likely sources or points of exposure. It is critical that all exposure points are identified so that the specification is accurate and expensive retro-fitting is avoided. The thoroughness of the risk assessment in initially identifying the areas where control is needed is fundamental to ensuring the installation of an effective LEV system. Regardless of whether the installation is likely to be simple or complex, discuss with supervisor who know his process, should work closely with any supervisor or lecturer to achieve an effective LEV design. Simple design process can be explained, which select the potential supplier who provided best LEV's. For supplier, should prepare design and quotation on exhaust the air contaminant from being exposed to operator. Then the supplier selected the author as well selected great design LEV's plus provided prototypes before deciding whether selected it or not. Figure 1 shows the design selected LEV's that satisfy properly which will collect the air that contains the contaminant (close to the source), make sure they are contained and taken away from people and clean the air and get rid the contaminant safely.

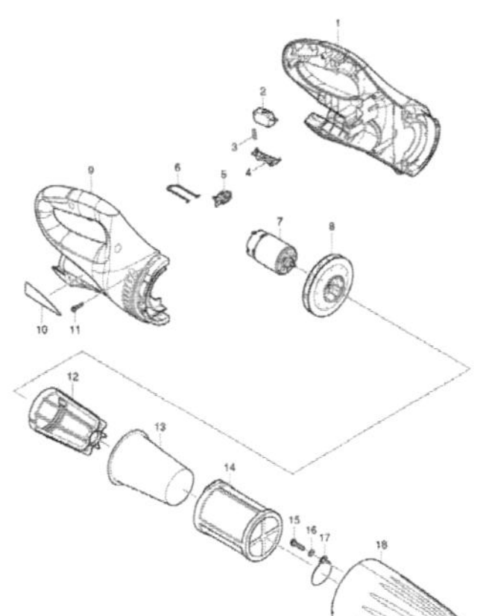

Figure 1. LEV System in Exploded View Spare Parts Model 10.8V LXT Cordless Air blower.

2.2. Experiment and Simulation on Design Selected LEV's

LEV is a ventilation system that collects and sucks out particles such as dusts, mists, gases, vapors or fumes out of work station, so that they can't be breathed in by occupants. To overcome this issue, SolidWorks will be implemented. Past studies SolidWorks techniques represent a very significant improvement of air ventilation systems. However, SolidWorks is just a tool in prediction model, which can lead to inaccuracy of predicting airflow due to problems with pre-processing, solver and post-processing with parameter from actual experimental results [10]. Yet, it is not possible to 100% accurately simulate airflow around a body [11]. These codes are simply models which are close to that of a real flow, but not an exact match. These require validation to help minimizing percentage error in SolidWorks methodology. Several strategies are needed to boost effectiveness of LEV in terms of predicting airflow in a geometry model [12]. The outcome of this research can be used as a benchmark or guideline for industries to help improving indoor air quality (IAQ). Figure 2 shows that the anemometer selected use for to do an experiment sample of duct or gas vapor produced by a laser which detect and measure the general duct or smoke velocity produced by laser inside whole chamber room that include operator who nearly exposed to it.

Figure 2. Mastech MS6252A Handheld LCD Digital Electronic Wind Speed Meter Anemometer Lock Function Digital Anemometer.

3. Results and Discussion

3.1. Bulk Microstructural Analysis

Given general duct velocity for smoke or fumes produced by laser 5-10 m/s, for air and carbon dioxide (CO2) which can be consider as general produced by laser in 5-10 m/s take as experiment for simulation as result in Figure 3 and Figure 5. Table 1 and Table 2 shows that to get how given designed made for LEV flow inside it and take velocity on air flow take in time to exhaust to outside surrounding workplace and then record in table below shows that after taking general air and carbon dioxide (CO2) velocity 5-10 m/s, being captured and velocity that air and carbon dioxide flow to be exhausted to outside surrounding workplace and take time for it. Then, proceed to identify the efficiency of LEV's to capture the all the duct or smoke for both air and carbon dioxide gases. Figure 4 and figure 6 shows that increases indicated velocity in graph by air and CO2 produced and captured and the gradient from both graph can be determined as efficiency of LEV's and then compare data given efficiency either satisfy for operator being safe from exposed hazard contaminant with at least 70- 90% efficiency.

Table 1. Air Flow Velocity Captured Given Air Dust Velocity Produced by Laser.

Air Velocity (m/s)	Air Flow Velocity Captured				
	Minimum Velocity (m/s)	Maximum Velocity (m/s)	Time (s)	Average Velocity (m/s)	Efficiency (%)
5	0.800	7.196	5.480	3.998	79.960
6	0.943	8.488	5.800	4.716	78.592
7	1.186	10.672	6.320	5.929	84.700
8	1.204	10.839	6.760	6.022	75.269
9	1.380	12.422	7.600	6.901	76.678
10	1.533	13.800	9.280	7.667	76.665
Air Duct/Smoke					

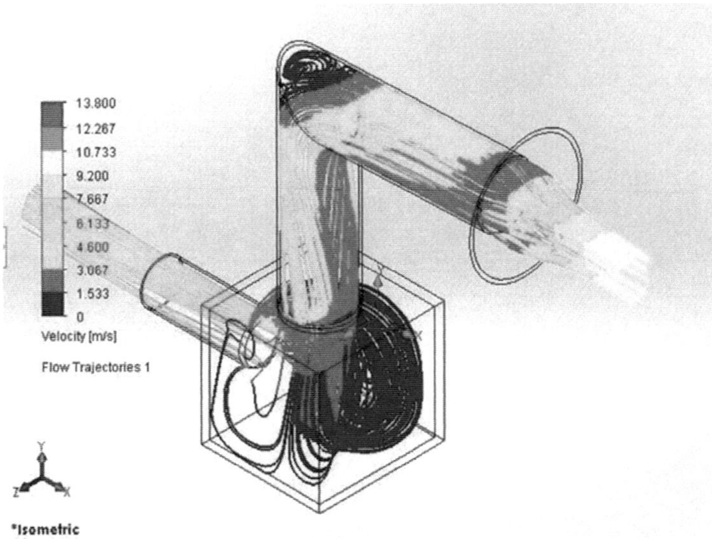

Figure 3. Air flow Simulation in LEV System Designed with indicates the maximum and minimum air velocity captured by LEV's.

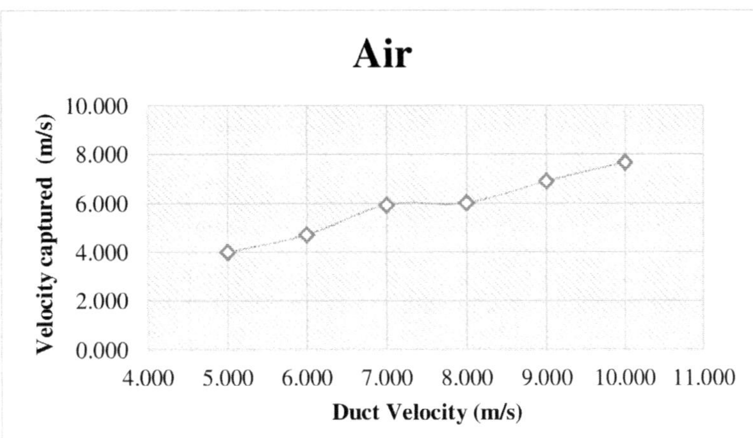

Figure 4. Graph of Air Duct velocity captured versus air duct velocity produced by laser.

IOP Conf. Series: Materials Science and Engineering **238** (2017) 012013 doi:10.1088/1757-899X/238/1/012013

Table 2. CO2 Flow Velocity Captured Given CO2 Dust Velocity Produced by Laser.

Carbon Dioxide (CO2) Velocity (m/s)	Carbon Dioxide (CO2) Flow Velocity Captured				
	Minimum Velocity (m/s)	Maximum Velocity (m/s)	Time (s)	Average Velocity (m/s)	Efficiency (%)
5	0.735	6.613	3.720	3.674	73.480
6	0.922	8.302	4.080	4.612	76.867
7	1.092	9.831	5.400	5.462	78.021
8	1.272	11.447	6.880	6.360	79.494
9	1.435	12.918	7.000	7.177	79.739
10	1.651	14.858	8.720	8.255	82.545
Air Duct/Smoke					

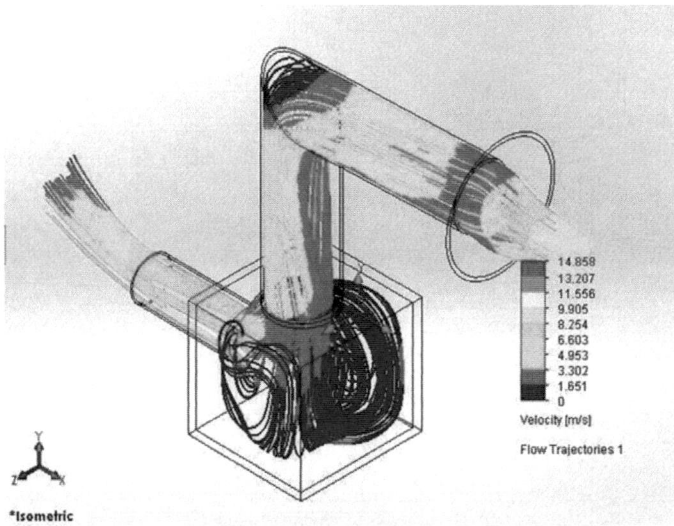

Figure 5. Carbon Dioxide Flow Simulation in LEV System Designed.

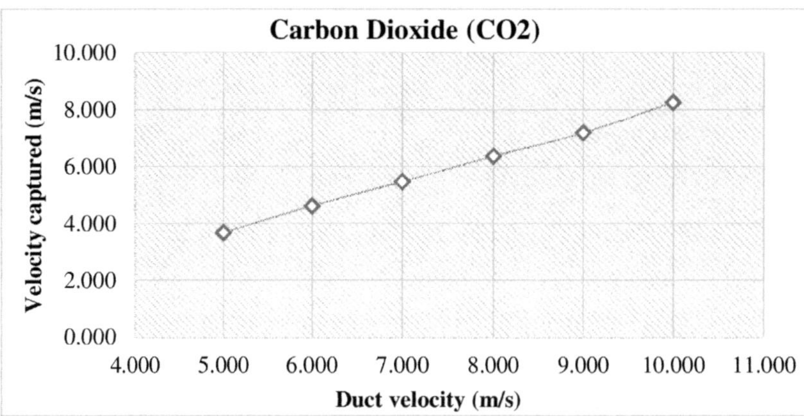

Figure 6. Graph of Carbon Dioxide (CO2) Dust velocity captured versus CO2 duct velocity produced by laser.

4. Conclusion

As conclusion, design the cordless local exhaust ventilation system and it with successfully installed at chamber room of laser cutting machine. LEV's fabricated with rated voltage D.C 10.8V and 1.5 ampere. Its capacity 600 ml, continuously use limit approximately 12-15 minute, overall length LEV's fabricated is 966 mm with net weight 0.88 kg and maximum airflow is 1.3 meter cubic per minute. Stimulate the air flow inside chamber room of laser cutting machine with the ventilation system that designed and fabricated overall result get 2 main gas vapor which air and carbon dioxide. For air gas which experimented by using anemometer, general duct velocity that produce is same with other gas produce, carbon dioxide which 5 m/s until 10 m/s. Overall result for 5 m/s and 10 m/s as minimum and maximum duct velocity produce for both air and carbon dioxide. The air gas flow velocity that captured by LEV's fabricated, 3.998 m/s average velocity captured from 5 m/s duct velocity which it efficiency of 79.960% and 7.667 m/s average velocity captured from 10 m/s duct velocity with efficiency of 76.665%. For carbon dioxide gas flow velocity that captured by levs fabricated, 3.674 m/s average velocities captured from 5 m/s duct velocity which it efficiency of 73.480% and 8.255 m/s average velocity captured from 10 m/s duct velocity with efficiency of 82.545%.

Acknowledgement

All experimental works were done in Joining and Welding lab, Faculty of Mechanical Engineering, UMP.

References

[1] Yu H, Cheng W, Wu L, Wang H and Xie Y, Mechanisms of dust diffuse pollution under forced-exhaust ventilation in fully-mechanized excavation faces by CFD-DEM *Powder Technology* 2017 317 31-47

[2] Ahmed A Q, Gao S, and Kareem A K, Energy saving and indoor thermal comfort evaluation using a novel local exhaust ventilation system for office rooms *Applied Thermal Engineering* 2017 110 821-834

[3] Arghand T, Taghi K, Hazim A and Mathias C, An experimental investigation of the flow and comfort parameters for under-floor, confluent jets and mixing ventilation systems in an open-plan office *Building and Environment* 2015 92 48-60

[4] Croitoru, C, Bode F, Nastase I, Sandu M, Vasilescu A and Tacutu L, General Ventilation System Optimization Study for Environment Improvement of Sludge Dewatering Area from a Wastewater Treatment Plant *Energy Procedia* 2017. 112 640-649

[5] Zhao J. and Cheng P, A lattice Boltzmann method for simulating laser cutting of thin metal plates *International Journal of Heat and Mass Transfer* 2017 110 94-103

[6] Shin J S, Oh S Y, Park, H, Chung C M, Seon S, Kim T-S, Lee L, Choi B-S and Moon J-K, High-speed fiber laser cutting of thick stainless steel for dismantling tasks. Optics & Laser Technology, 2017. 94: p. 244-247

[7] Guerra A J, Farjas J, and Ciurana J, Fibre laser cutting of polycaprolactone sheet for stents manufacturing: A feasibility study *Optics & Laser Technology* 2017 95 113-123

[8] Schleier M, Adelmann B and Neumeier B, Burr formation detector for fiber laser cutting based on a photodiode sensor system *Optics & Laser Technology* 2017 96 13-17

[9] Samarjy R S M and Kaplan A F H, Using laser cutting as a source of molten droplets for additive manufacturing: A new recycling technique *Materials & Design* 2017 125 76-84

[10] Hunt G R and Ingham D B, 03 O 03 A 3-dimensional axisymmetric model of the air flow pattern created by a local exhaust ventilation system reinforced by a turbulent radial jet flow *Journal of Aerosol Science* 199. 24 S15-S16

[11] Varghese R J, Kishore V R V, Akram M, Yoon Y and Kuma Burning velocities of DME (dimethyl ether)-air premixed flames at elevated temperatures *Energy*, 2017 126 34-41

[12] Putra R N and Ajiwiguna T A, Influence of Air Temperature and Velocity for Drying Process *Procedia Engineering* 2017 170 516-519

IOP Conf. Series: Materials Science and Engineering **238** (2017) 012014 doi:10.1088/1757-899X/238/1/012014

The effect of multiple reflow on intermetallic layer of Sn-4.0AgCu/Cu by using microwave and reflow soldering

Jaya I Rasbudin, M A Rabiatul Adawiyah and O Saliza Azlina

Faculty of Mechanical and Manufacturing Engineering, University of Tun Hussein Onn Malaysia, 86400 Parit Raja, Johor, Malaysia

E-mail: salizaz@uthm.edu.my

Abstract. Soldering method is a general trend in demand that can be used to analyse different process parameter and their relationship in faster and easier way. Different soldering methods will influence the growth of intermetallic layer. This study investigates the effect of multiple reflow on intermetallic layer of Sn-4.0Ag-0.5Cu with diameter 700 µm solder size and Cu substrate by using microwave soldering. As comparison, reflow soldering was conducted. The material characteristic was characterised by using optical microscope, emission scanning (SEM) and energy dispersive X-ray analysis (EDX). The results revealed that only Cu6Sn5 intermetallic compound (IMC) was formed on solder joint. Meanwhile, Cu3Sn layer was detected after second and third reflow. The results also indicated that the soldering method could influence the IMC growth, where the microwave soldering produced a thinner IMC layer than reflow soldering. Besides, the morphology of IMC was observed as smaller circle bumps and scallop shapes. Moreover, the grain sizes of IMC for reflow soldering presented a larger, denser and rounder compared to microwave soldering.

1. Introduction

The formation of intermetallic compound (IMC) layer between lead-free solders and substrate is necessary for solder joint. The solder joint between the electronic device and substrate is a main interconnection for the performance and reliability of an electronic system. The IMC formation take place when there is a metallurgical reaction between a molten solder and metallization on the substrate [5, 13]. Basically, two types of IMCs which are Ag_3Sn and Cu6Sn5 can be formed at interface. Large and excessive IMCs layer are quite brittle, where it can reduce the reliability of solder joint. Jen et al (2011) reported that, the brittleness of IMC caused the strong stress concentration effect during thermal cycle, and the crack propagation could easily occur near the IMC layers [6]. Subsequently, the growth of IMC could be influenced by type and size of solder, surface finish, bonding temperature and time, as well as soldering method.

Soldering is one of the important processes to melt the metals for the permanent joining of materials. The understanding of soldering process is necessary to achieve good performance of printed circuit board (PCB) manufacturing process. In electronic industrial processes, soldering method is a common trend demand that can be used to analyze difference process parameters. One of the soldering methods is using conventional furnaces such as blast furnace, induction furnace, electric arc furnace and crucible furnace [1, 12]. However, the conventional method consumes a large amount of energy and also material and energy losses [12]. In order to surpass the disadvantages of conventional furnaces, microwave processing was introduced because of their advantages such as high heating rate, shorter processing time, and low power consumption and less environmental hazard. According to

Moon et al (2005), the growth of IMC can be affected by the peak temperature and holding time in the reflow process, where a faster heating and shorter holding time can reduce the thickness of IMC layer as well as its grain size [11].

Furthermore, microwave processing also produces large amount of heat, where it can generate losses material via the volumetric. Compared to conventional furnaces, the material is melted or heated throughout external heat source as well as radiative transfer [7]. During microwave heating, it can cause sparking of metallic materials and most metals are known to reflect microwave. In this process, susceptor or cladding is used as microwave absorbing materials, where it can develop heat energy to the bulk material as well as relatively lower heating efficiency [1, 2]. Several absorbing materials such as charcoal, silicon carbide, graphite and iron powder are used in microwave soldering [1, 2, 10, 15, 16]. This paper presents the results of an experimental investigation to the effect of multiple reflow on intermetallic layer in term of thickness, type and morphology of Sn-4.0Ag-0.5Cu/Cu by using microwave soldering and reflow soldering.

2. Experimental Procedure

2.1. Material selection

In this study, the double type copper with polymer (FR-4) was selected as the substrate for joining with dimension of 40 mm x 50 mm x 1.2 mm. Solder alloy of Sn-4.0Ag-0.5Cu (SAC405) with a diameter of 700 μm was prepared for soldering. Prior to soldering, a thin layer no-clean flux was applied. Then, the solder balls were manually arranged in several rows onto the Cu substrate. The no-clean flux worked to avoid oxidation from the Cu substrate and improve the wetting of solder ball during soldering process.

2.2. Joining process

In this study, microwave is used as a soldering technique and reflow soldering for comparison. By using microwave soldering, the samples were put in the crucible alumina boat which contained iron powder, where it acted as an absorber material. Besides that, the usage of crucible could improve heating efficiency, where it can generate microwaves heat very well. The crucible that contained the samples was closed with the lid/top cover to prevent the microwave energy reflection to the Cu substrate and sent into a microwave with frequency of 2.45GHz and 800W for 60 seconds. For reflow soldering, the conventional oven was used at the reflowing temperatures of 230°C for 20 minutes to form a joint at the SAC405/Cu interface. After first reflow, the samples were exposed to second and third reflow (multiple reflow) and cooled under room temperature before they were cleaned with acetone and dried for both soldering method.

2.3. Characterization methods

Two methods of the cross-sectional and top surface were prepared for metallographic characterization. The characterizations of IMC layer on the thickness and morphology were examined using the scanning electron microscope (SEM) and image analyzer (IA). Energy dispersive x-ray (EDX) was used to identify the element of IMC composition.

3. Results and Discussion

The SEM micrographs in Figure 1 and Figure 2 show the IMC growth morphologies of SAC405/Cu after multiple reflow for both reflow soldering and microwave soldering, respectively. During reflow process, the solder ball of SAC405 melted into contact with Cu substrate and then Sn in the molten solder reacted with Cu to form IMC layer at interface. As-reflowed joint, the IMC morphology of Cu_6Sn_5 was formed at the SAC405/Cu interface as shown in Figure 1 (a) and Figure 2(a). The formation of Cu_6Sn_5 occurred because of the Cu dissolution, followed by a chemical reaction. Chemical reaction occurred between solder and conductor metal such as Cu substrate and IMC nucleate, and IMC grew at the solder/substrate interfaces [8, 19]. In the case of the second and third reflow, two types of IMCs which were Cu_6Sn_5 and Cu_3Sn appeared at interface. When the molten solder was exposed long enough with temperature or time, a new layer of Cu_3Sn was formed between Cu and Cu_6Sn_5. Tu *et al* (2017) and Hua *et al* (2014) reported that, Cu_3Sn was formed by diffusion reaction between Cu atom and Sn atom from the solder [4, 18]. These morphologies can be classified into two majors; scallop and circle bump shapes. It is obviously seen that the grain sizes of IMC become bigger and denser with increasing number of reflow cycles. Meanwhile, the grain size of microwave soldering technique is smaller than reflow soldering as presented in Figure 1 and Figure 2.

Figure 1. Top surface images of SAC405/Cu with solder size Ø700 μm by using reflow soldering: (a) first reflow, (b) second reflow and (c) third reflow.

Figure 2. Top surface images of SAC405/Cu with solder size Ø700 μm by using microwave soldering: (a) first reflow, (b) second reflow and (c) third reflow.

The cross-sectional morphology of SAC405/Cu solder joints with different soldering method was presented in Figure 3 and Figure 4. The results showed that, Cu_6Sn_5 grew as scallop grains with rough interfacial morphology between Cu substrate and the solder. Whereas, a thin layer of Cu_3Sn was found between Cu_6Sn_5 and Cu substrate. From the cross-sectional images, it is clearly seen that the IMC thickness is greater with the reflow times. This phenomenon might happen because of the Cu concentration in solder balls. When the reflow process exposed to third reflow, Cu concentration in solder was increased and thick IMC layer was produced. If the IMC layer became thicker, it could reduce the reliability of solder joint because the behavior of IMC is brittle [14, 17]. The thickness of IMC layers corresponding to first reflow, second reflow and third reflow were 3.77 μm, 5.73 μm and 6.07 μm for reflow soldering, respectively. The thickness of IMC by using microwave soldering for first reflow, second reflow and third reflow were 2.18 μm, 3.79 μm and 5.35 μm, respectively. These results proved that the thickness of IMC layer can influence the solder joint reliability after multiple reflow cycle.

Comparing the two results of IMC in term of thickness and grain sizes by using the reflow soldering and microwave soldering after multiple reflow process, it was clearly seen that the growth of IMC morphologies by using microwave soldering was thinner and smaller than the IMC results of reflow soldering. Based on the results, it was reasonable to mention the IMC formation on SAC405/Cu grew faster by using reflow soldering. This situation might happen because of the longer reaction time during reflow process and make it accelerates of the IMC growth. As compared to IMC formation by using microwave soldering, the IMC grew is slower and produced a thin and small grain sizes at interface because of shorter processing time during reflow process. These results are supported by Srinath *et al* (2011) and Moon *et al* (2005) [11, 15]. These results are reasonable to suggest that by using the microwave as reflow process, it can reduce the growth of IMC formation at interface and enhance the reliability of solder joint.

Moreover, the IMCs of Cu_6Sn_5 and Cu_3Sn were not only detected at interface, but Ag_3Sn was also found in the solder matrix of solder joint as shown in Figure 3 and Figure 4. Besides, Ag_3Sn was observed at the SAC405 bulk solder. Figure 5 show the top surface view of the morphology with deep etching process. From the figures, the IMC of Ag_3Sn was formed as long and needle like as well as plate-like, as seen in Figure 5(a) and Figure 5(b). Chung and Tai (2004) mentioned that large Ag_3Sn was initially existed in the solder bulk and could be observed at the interface onto or next to the Cu_6Sn_5 IMC [3]. Furthermore, many small particles or nanoparticles of Ag_3Sn were observed on the Cu_6Sn_5 grains with white sports as shown in Figure 1 and Figure 2. This happened during solidification process, where the Ag_3Sn nanoparticles could reduce the interfacial energy and retard the growth of the IMC layer. Similar observation was discovered by Liu *et al* (2010) [9]. However, the size of this IMC would not change or grow with increasing number of reflow time. Subsequently, large Ag_3Sn can be affected the reliability of solder joint where it can lead to crack initiation and propagation. Furthermore, large Ag_3Sn also caused solid dissolution and precipitation hardening. As a result, the solder joint became weak [8].

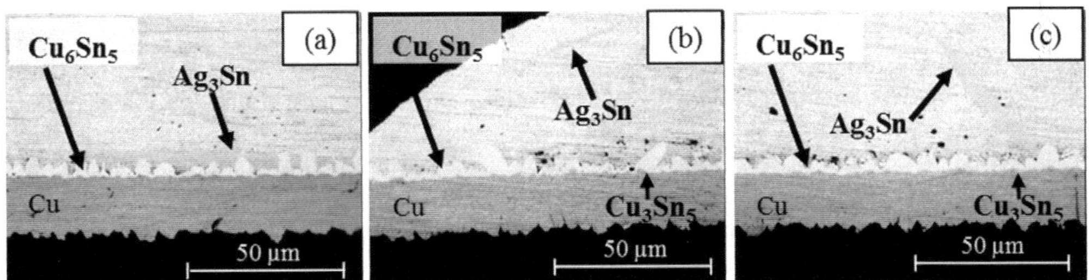

Figure 3. Optical microscope images of cross-sectional view for SAC405/Cu with solder size Ø700 µm by using reflow soldering: (a) first reflow, (b) second reflow and (c) third reflow.

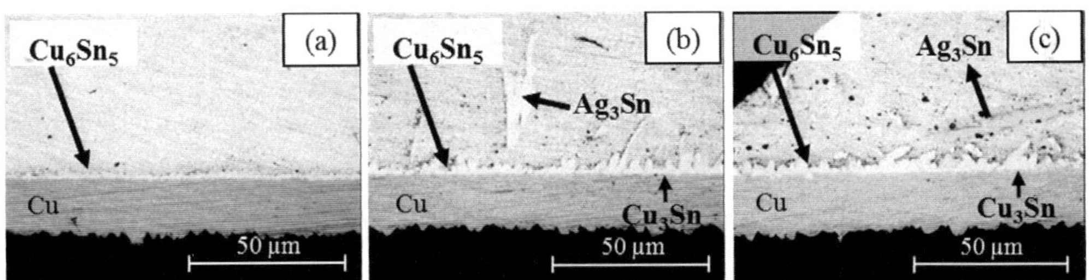

Figure 4. Optical microscope images of cross-sectional view for SAC405/Cu with solder size Ø700 µm by using microwave soldering: (a) first reflow, (b) second reflow and (c) third reflow.

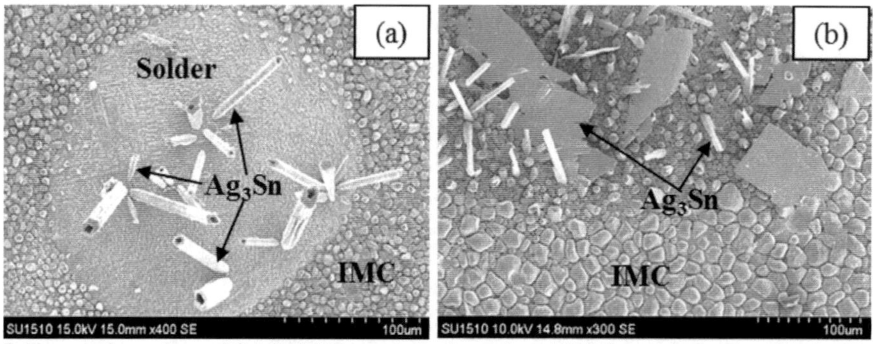

Figure 5. Top surface view of SAC405/Cu with 3-dimentional IMC morphologies: (a) Ag_3Sn IMC at bulk solder and (b) Ag_3Sn at interface with Cu_6Sn_5 IMC.

4. Conclusion

The lead-free solders of SAC405/Cu that were joined with Cu substrate to form IMC layer by using microwave and reflow soldering was studied. The major conclusions of the study can be summarized as:

i. For both soldering method, only Cu_6Sn_5 IMC was observed at interface as scallop and circle bump shapes. After exposed to second and third reflow, new layer IMC of Cu_3Sn was formed between Cu_6Sn_5 and Cu substrate. Ag_3Sn was also formed as long or needle like and plate-like.

ii. The IMC thickness and grain sizes became thicker, larger and denser with increasing number of reflow cycle, for both soldering method.

iii. Microwave soldering method represented thinner IMC layer and smaller grain sizes compared to reflow soldering.

Acknowledgement

The authors would like to thank the University of Tun Hussein Onn Malaysia (UTHM) and research facilities given by Faculty of Mechanical and Manufacturing Engineering, UTHM.

References

[1] Chandrasekaran S, Basak T and Ramanathan S 2011 Experimental and theoretical investigation on microwave melting of metals *J. Mater. Process. Technol* 211 482–87

[2] Chandrasekaran S, Basak T and Srinivasan R 2013 Microwave heating characteristics of graphite based powder mixtures *Int. Commun. Heat Mass Transf.* 48 22–7

[3] Chung C K and Tai S F 2004 Evolution of Ag3Sn during reflow soldering Proc. 9th Therm.Thermomechanical Phenom *Electron. Syst. Conf.* 2 116–20

[4] Hua X, Li Y and Min Z 2014 Interfacial reaction and IMC growth between Bi-containing Sn0.7Cu solders and Cu substrate during soldering and aging *J. Alloy. Compd.* 582 341–47

[5] Huang M L and Yang F 2015 Solder size effect on early stage interfacial Intermetallic compound evolution in wetting reaction of Sn3.0Ag0.5Cu/ENEPIG joints *J. Mater. Sci. technol.* 31 252–56

[6] Jen Y-M, Chiou Y-C and Yu C-L 2011 Fracture mechanics study on the intermetallic compound cracks for the solder joints of electronic packages *Engi. Fail. Anal.* 18 797–810

[7] Jones D A, Lelyveld T P, Mavrofidis S W, Kingman S W and Miles N J 2002 Microwave heating applications in environmental engineering—a review Resour *Conserv. Recycl.* 34 75–90

[8] Lee L M and Mohamad A A 2013 Interfacial reaction of Sn-Ag-Cu lead-free solder alloy on Cu: A review *Adv. Mater. Sci. Eng.* 2013 1–11

[9] Liu X, Huang M, Zhou Y, Wu C M L and Wang L 2010 The adsorption of Ag3Sn nano-particles on Cu-Sn intermetallic compounds of Sn-3Ag-0.5Cu/Cu during soldering *J. Alloy. Compd.* 492 433–38

[10] Mohamad Faizal S and Saliza Azlina O 2016 Intermetallic compound formation on lead-free solders by using microwave energy *Proc. AIP Conf.* 1-6
[11] Moon K-S, Li Y, Xu J and Wong C P 2005 Lead-free interconnect technique by using variable frequency microwave *J. Electron. Mater.* 34 1081–88
[12] Moore A, Schechter D and Morrow M 2003 Method and apparatus for melting metals U S Patent 0089481 1 1–3
[13] Sona M and Prabhu K N 2013 Review on microstructure evolution in Sn-Ag-Cu solders and its effect on mechanical integrity of solder joints *J. Materi. Sc.: Mater. Electron.* 24 3149–69
[14] Sona M and Prabhu K N 2014 The effect of reflow time on reactive wetting, evolution of interfacial IMCs and shear strength of eutectic Sn-Cu solder alloy *J. Mater. Sci.: Mater. Electron.* 25 1446–55
[15] Srinath M S, Sharma A K and Kumar P 2011 A new approach to joining of bulk copper using microwave energy *Mater. Des.* 32 2685–94
[16] Srinath M S, Sharma A K and Kumar P 2011 Investigation on microstructural and mechanical properties of microwave processed dissimilar joints *J. Manuf. Process.* 13 141–46
[17] Tseng C F, Jill Lee C and Duh J G 2013 Roles of Cu in Pb-free solders jointed with electroless Ni(P) plating *Mater. Sci. Engi. A* 574 60–7
[18] Tu X, Yi D, Wu J and Wang B 2017 Influence of Ce addition on Sn-3.0Ag-0.5Cu solder joints: Thermal behavior, microstructure and mechanical properties *J. Alloy. Compd.* 698 317–28
[19] Xian J W, Belyakov S A, Ollivier M, Nogita K, Yasuda H and Gourlay C M 2017 Cu6Sn5 crystal growth mechanisms during solidification of electronic interconnections. *Acta Mater.* 126 540–51

Gas Metal Arc Welding Parameters Effect on Properties of Tailored Orbital Weld of SS304 and BS1387

M N Ayof, N I S Hussein* and M Z Mohd Noh
Faculty of Manufacturing Engineering, Universiti Teknikal Malaysia Melaka, Hang Tuah Jaya, 76100 Durian Tunggal, Melaka, Malaysia.

*E-mail: izan@utem.edu.my

Abstract. Dissimilar material pipes in a power plant boiler water piping system are used to transmit water at various temperatures, either in extremely high temperature water or room temperature water. In this study, tailored orbital welding of dissimilar material of Stainless Steel (SS) 304 and British Steel (BS) 1387 were performed by Gas Metal Arc Welding (GMAW) with automated fixed nozzle-rotational jig. This study focused on GMAW parameters variation effects on mechanical properties of SS304 and BS1387 dissimilar material tailored orbital welding. The weldment quality was tested by performing non-destructive dye penetrant test. The tensile strength and microhardness were studied to verify the influence of welding parameters variations. Design of Experiment (DOE) was employed to generate process parameter using Response Surface Methodology (RSM) method. Welding parameters that were arc current, arc voltage and travel speed as input response, whilst, tensile strength and microhardness as output response. Results from non-destructive test showed no major defect occurred. The tensile strength and microhardness increased when arc current and voltage increased and travel speed decreased. Microhardness at weldment was higher than base material.

1. Introduction

Welding is a versatile joining process that is applicable to almost all types of materials. It is one of the permanent joining processes that produce coalescence of the material by heating workpiece to the melting temperature with or without the existence of pressure or by the application of pressure itself and with or without the use of filler material for metal or non-metallic materials. Welding technique has been widely used in various industries such as automotive, oil and gas, aerospace, and many others. There are various types of welding and it differs according to the heat source, process and type of welded material such as, shielded metal arc welding (SMAW), gas metal arc welding (GMAW), gas tungsten arc welding (GTAW) and laser welding. In many industries, welding plays an important role in reducing the production cycle time, thus reducing the delivery time [1]. Orbital welding is a joining process of tubes of similar thickness. It was one of the major improvements in pipe welding technology since 1980s. The orbital arc welding produces high quality of welded seams and has good repeatability. It is an expansion to manual welding process as it increases speed and ensures repeatability [1].

Tailored orbital (TO) is welded tubular product made of tube from different materials of different thickness or coatings to component or other tubes, depending on the application in taking advantages in cost, weight and function. If the material joint is different to each other, it is also known as dissimilar metal joint (DMJ). The advantages of TO technology are it can help to improve the production processes as they reduce material requirements and steps, reduce overall part weight, improve functionality, reduce costs and allow greater forming freedom [2].

One of the most common industries that apply tailored orbital welding (TOW) is in power plant boiler system as an example shown in Figure 1. In power plant industry, there are various components or systems operating at different service conditions hence appropriate materials are used for high temperatures or low temperatures condition [3-7]. Materials used are selected depending on the requirement of working environment; high temperature require high properties steel such as stainless steel (SS) and low properties steel such as carbon steel (CS) for low temperature is necessary [3]. Due to economic pressure, the application of TOW is essential in power plant because it is able to minimize requirement on high performance steel and maximize steel capabilities at appropriate area [8][6-7].

Figure 1. Example of TOW application in power plant [9]

There are various steel combinations in dissimilar material such as between CS and SS, ferritic steel and SS, and martensitic steel and austenitic stainless steel, and it depends on its application. CS and SS is the most common steel combination in TOW process applied at power plant due to economic and technical reason. Akbari and Sattari-Far [3] performed dissimilar material of CS A106B and SS A240-TP304 to carry out failure analysis on instance steam generators of power plants, Hajiannia et al. [4] used CS A 335 and SS 347 to replicate the heat exchangers of power plant and Lee et. al. [6] utilized CS SPPS 42 and SS304 as both materials were commonly found in power plant due to economic and technical reasons.

In TOW, welding is the most common method used as heat source. GTAW was used by Akbari and Sattari-Far [3], Hajiannia et al. [4] and Lee et al. [6] to perform TOW. The GTAW is able to produce high quality weld. However, it requires high skill welder to control both hands and increased agility. In order to attract the interest of power plant industries, GMAW was used as heat source for this research study. The study of TO welding using GMAW was performed using SS wire as filler material. The study was carried out by automatic feeding of a continuous consumable electrode and fixed nozzle-rotational jig. Thus, dependence on high skill welder was replaced by the usage of fixed nozzle-rotational jig.

The objectives of the study are to correlate the effect of different GMAW parameters on the tensile strength and microhardness variations of the TOW of SS304 and BS1387.

2. Experimental Methods

2.1. Material

The materials used in this experiment were two different types of pipe material with the same outside diameter (OD), different inside diameter (ID) and different thickness. The pipe materials used were SS304 and BS1387. The SS304 had 60.0mm OD, 56.40mm ID and 3.60mm thickness. Meanwhile, BS1387 had 60.00mm OD, 57.00mm ID and 3.00mm thickness. The filler material used in this experiment was ER 308L SS with 1.2 mm diameter.

IOP Conf. Series: Materials Science and Engineering **238** (2017) 012015 doi:10.1088/1757-899X/238/1/012015

2.2. Design of Experiment

The welding parameters were generated by using RSM; Box Behnken function in the Design Expert software. The three input variables or factors were welding current (A), welding voltage (V) and travel speed (RPM), and the two output variables or response, were tensile strength (MPa) and Vickers microhardness (HV). Each factor was subjected to two levels which were low and high levels. Table 1 shows the level of factors for the GMAW processes. By having three input variables and two output variables with five centre points per block, the total runs generated were 17 runs, or set of parameters.

Table 1. Levels of parameters factors for GMAW processes

Input variables (Factors)	Units	Low Level (-1)	High Level (+1)
Arc Current	A	150	170
Arc Voltage	V	17	21.5
Travel Speed	RPM	2	4

2.3. Welding Process

Pipes were held on the rotational jig with 90° or perpendicular to the welding nozzle and with 20.0 mm distance. The nozzle set on the centre of weld line. The shielding gas that was used in this experiment was a mix of 80% argon and 20 % CO_2. Rotational jig was used in this experiment to hold and rotate the pipe to perform tailored orbital welding. The rotational jig used was ANYEKE welding positioned APL-100 model. The rotational jig was connected to the welding machine used. The welding machine used was Fronius TransSynergic 4000 GMAW machine with VR 4000 wire feed system.

2.4. Dye Penetrant Test

The dye penetrant test was conducted according to AWS D1, with the following steps; pre-cleaning the surface using the Cleaner to clean the surface and make the surface smooth enough to wipe off the penetrant without leaving residue. Then, applying the penetrant, with light layer sprayed on the pipe with around 50.0 mm distance. The penetrant was let to permeate into crack and voids if any for 30 minutes. After 30 minutes, the penetrant was removed using cleaner and cloth. Then, applying the developer with light layer sprayed on the pipe with distance around 50.0 mm. The developer was dried for 30 minutes. Then, the pipe was evaluated and checked for any defects appeared.

2.5. Tensile Testing and Microhardness Measurement

For tensile test, these pipes were cut into dimensions in accordance with ASTM E8M-4 tension test specimens for large-diameter tubular products. AG-I Shimadzu Universal Testing Machine was employed.

For microhardness test, the sample was cut and the test was conducted in accordance with ASTM E92 for Vickers test. The microhardness test was performed by using Mitutoyo HR-523 machine, Vickers Microhardness Tester. The microhardness measurement was carried out by applying a load of 0.1 kgf with 15 sec dwell time.

3. Results and Discussion

3.1. Macrostructure

From the 17 samples, there was no major defect on the surface of the weldment. However, there are mostly series of red dot at the start-end of the weldment. Figure 2 shows sample 1 (160 A, 19.25 V and 3 RPM) with red dots at the start-end weldment. Dwivedi and Sharan [10] claimed that the red dot appearance indicates pits or porosity and series of red dots indicates crack or cold shuts. Based on the dye penetrant results, some samples had series of red dots while some had single red dot at start-end point. This indicates porosity and crack had occurred during the welding process, due to lack of fusion or inadequate weld at start-end point as supported by Siegel [11]. However, no significant porosity or crack was observed at the cross section of the weldment (Figure 3).

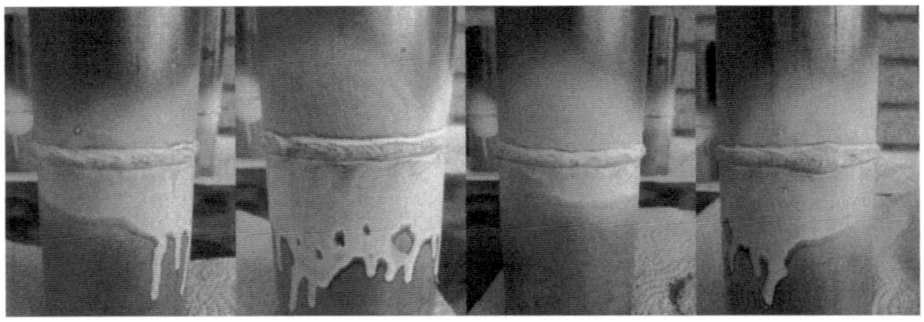

Figure 2. Sample no.1 of TOW with 160 A, 19.25 V and 3 RPM as parameter shows red dots at the start-end.

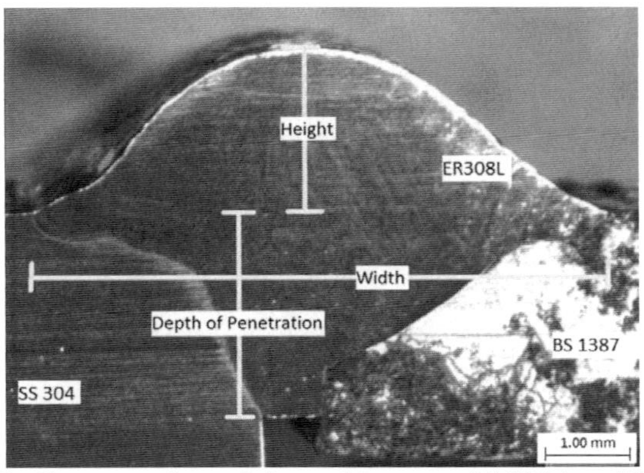

Figure 3. Cross section of the weldment of SS304 and BS1387.

3.2. Base Material Characterization

Base materials were charactierized by using tensile test and microhardness test to determine the tensile properties and microhardness values. The results were used as the benchmark for the weldment results later. Table 2 shows the tensile properties and microhardness values for SS304 and BS1387. The SS304 had 683.785 MPa tensile strength and 217.6 HV microhardness. Meanwhile, BS1387 had 390.766 MPa and 187.1 HV microhardness. However, BS1387 had lower elasticity with 5160.36 MPa Young's Modulus as compared to SS304 which had 5364.67 MPa Young's Modulus.

Table 2. Tensile Properties and microhardness of SS304 and BS1387 base material

Base Material	Tensile Properties		Microhardness
	Tensile Strength (MPa)	Young's Modulus (MPa)	Vickers (HV)
SS304	683.785	5364.67	217.6
BS1387	390.766	5160.36	187.1

3.3. Process-properties Relationship

Response Surface Methodology (RSM) was used to analyse the result obtained from the welding process and to understand the relationship between welding parameters with mechanical properties of TOW. The factors were welding current, welding voltage and travel speed, while responses were tensile strength. The design also analysed which factor influenced respond the most. All analysis was done by using Design Expert software. From this experiment, the highest tensile strength value was 351.35 MPa from sample 7. Meanwhile, the lowest tensile value was 249.79 MPa from sample 11. The highest microhardness value of weldment was 249.32 HV from sample 12, and the lowest microhardness value was 178.46 HV from sample 14.

3.4. Tensile Strength

3.4.1. Polynomial Equation for RSM Tensile Strength Model

In Design Expert 7 software, the appropriate polynomial equations were determined. It represented the relationship between the input parameters and the output response. It was done by carrying out SMSS and 'lack of fit' test. Both analyses suggested the quadratic equations to model and calculate the relationship between input parameters and micro results. The quadratic equation was the most significant with 0.0209 p-value for SMSS and 0.7187 p-value for lack of fit test.

3.4.2. ANOVA Analysis of the Response Surface Quadratic Model for Tensile Strength

ANOVA analysis for the quadratic model for tensile is shown in Table 3. The "Model F Value" of 4.89 implied that the model was significant. There was only 2.41 % chance that a "Model F-value" this large could occur due to noise. This implied that the model did represent the data within the required 90% confidence interval. The most significant was model term C (travel speed) with p-value of 0.0365. However, model term A (arc current) and model term B (arc voltage) were not significant with p-value 0.7598 and 0.4167, respectively. Model term A and B were included in the analysis for keeping the model hierarchy because interaction model term AB and C^2 was significant with p-value of 0.0060 and 0.0069 respectively.

Table 3. ANOVA analysis of the quadratic model for tensile strength

Source	Sum of Squares	Df	Mean Square	F Value	p-value Prob > F	
Model	13224.97	9	1469.44	4.89	0.0241	Significant
A-Arc Current	30.38	1	30.38	0.10	0.7598	
B-Arc Voltage	223.94	1	223.94	0.74	0.4167	
C-Travel Speed	2001.12	1	2001.12	6.66	0.0365	
AB	4556.25	1	4556.25	15.16	0.0060	
AC	18.69	1	18.69	0.06	0.8103	
BC	677.27	1	677.27	2.25	0.1771	
A^2	479.86	1	479.86	1.60	0.2469	
B^2	1079.09	1	1079.09	3.59	0.1000	
C^2	4284.48	1	4284.48	14.25	0.0069	
Residual	2104.35	7	300.62			
Lack of Fit	549.49	3	183.16	0.47	0.7187	not significant
Pure Error	1554.86	4	388.71			
Cor Total	15329.32	16				
				R^2	0.8627	
				Adj. R^2	0.6862	
				Pred. R^2	0.2680	
				Adeq. R^2	8.183	

The accuracy of this model was also supported by the lack of fit analysis. The "Lack of Fit F-value" was 0.47 and implied that the Lack of Fit was not significant relative to the pure error. There was 71.87% chance that a "Lack of Fit F-value" this large could occur due to noise. The non-significant lack of fit was good because it made the model fit. The value coefficient of determination, R^2 was 0.8627 hence, the value of correlation coefficient, r was 0.9288 which was higher than 0.8. A correlation greater than 0.8 was generally described as strong. The value of Pred. R^2 of 0.2680 was not close to the Adj. R^2 of 0.6862 as one might normally expected. This may indicate a large block effect or a possible problem with model and/or data. However, Adeq. Precision indicated the measured signal to noise ratio was desirable, which was greater than 4. The ratio was 8.183 thus indicated an adequate signal and the model can be used to navigate the design space.

There were three variables, welding current, voltage and travel speed. However, only travel speed was significant to the model but welding current and voltage was included in the model for keeping the model hierarchy. It was because the interaction between welding current, voltage and travel speed was significant.

3.4.3. Significant Factors Influencing Tensile Strength

The significant factors influencing tensile strength value were determined by using ANOVA analysis of the quadratic model for response – tensile strength, Table 3. Based on the p-value less than 0.05, travel speed (C), interaction between current and voltage (AB) and travel speed quadratic term (C^2) were the significant influencing factors of the tensile strength.

3.4.4. Effect of Travel Speed (C) on Tensile Strength

Based on Table 3, travel speed p-value was 0.0365, which was less than 0.05 and was classified as one of the significant factors influencing tensile strength. The travel speed main factor curve in Figure 4 shows that as the tensile strength decreased from 339.74 MPa to around 280 MPa, the travel speed increased from 2 RPM to 3 RPM. However, as the tensile strength increased after passing 3 RPM until maximum 4 RPM, with tensile strength value 308.12 MPa. Similar result was found by Talabi et al. [12], as they increased the travel speed, the tensile strength decreased and at certain speed and further increase in the travel speed, the tensile strength increased.

Effect of travel speed showed that the increased in travel speed decreased the tensile strength value. Similar result was found by Ampaiboon et al. [13] whereas the maximum UTS of welded joint was obtained by lowest travel speed and increasing the travel speed decreased the tensile strength gradually.

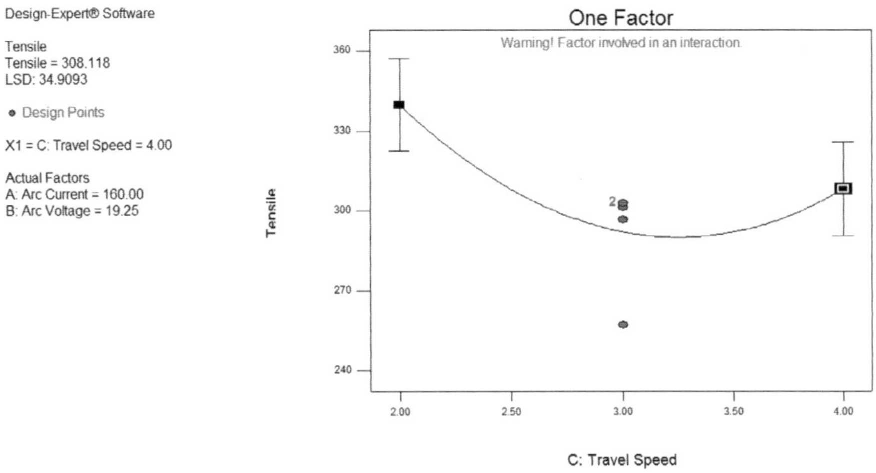

Figure 4. Behaviour of weldment tensile strength in response to variation of welding travel speed

3.4.5. Effect of Interaction between Arc Current and Arc Voltage (AB) on Tensile Strength

The ANOVA analysis revealed that the interaction between arc current and arc voltage (AB) is the most significant factor on tensile with p-value of 0.006. Figure 5 shows the interaction between arc current and arc voltage with respect to the weldment tensile strength. At low arc voltage, 17.0 V, the variation in arc current, increased the weldment tensile strength value, from 245.71 MPa to 317.11 MPa. However, at high arc voltage, 21.5 V, variation in arc current decreased the weldment tensile strength value from 323.79 MPa to 260.19 MPa. This indicated that the interaction between arc current and arc voltage significantly affected the weldment tensile strength value.

The effect of arc voltage on tensile strength is the increase in arc current which lead to the increase in tensile strength value. Similar result was found by Ampaiboon et al. [13] which focused on the effect of welding parameters on ultimate tensile strength. Maximum UTS of welded joint was obtained by highest voltage value. It is also supported by Ragu Nathan et al. [14] that the weldment is comparatively stronger due to strong carbide or nitride element formation which has limited solubility in ferrite and austenite.

Therefore, at low voltage, the increased in the arc current resulted in the tensile strength value to increase. However, at high voltage, variation in arc current significantly decreased the tensile strength value. Similar results were observed by Hussein et al. [15].This was because at high voltage produced wider, flatter and less penetration as compared to low voltage welding. As mentioned by Rohit and A.K. [16], the depth of penetration was at maximum for optimum arc voltage.

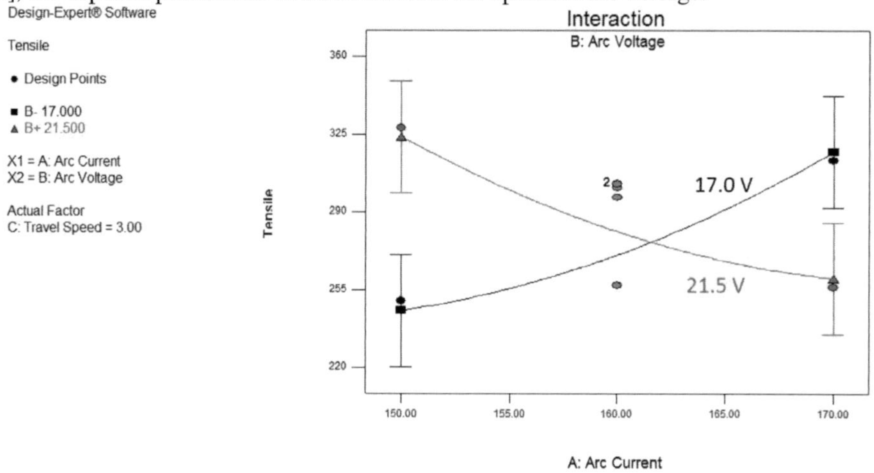

Figure 5. Interaction between arc current and arc voltage with respect to welding tensile strength

3.5. Microhardness

The significant factors influencing hardness were also determined by using ANOVA analysis of the quadratic model for response – microhardness. Based on the p-value less than 0.05, arc current (A), interaction between current and voltage (AB), interaction between current and travel speed (AC), interaction between voltage and travel speed (BC), arc voltage quadratic term (B^2), travel speed quadratic term (C^2) and interaction between current and voltage quadratic term (AB^2) were the significant influencing factors of the microhardness. As the arc current increased from 150 A to 170 A, microhardness of the weld decreased from 253.75 HV to 213.97 HV. Similar result was found by Bodude and Momohjimoh [17] as they increased the current value with constant voltage, the hardness decreased.

Increased in arc voltage from 17 V to 21.5 V led to insignificant change of weldment microhardness from 213 HV to only 214 HV. However, the quadratic term of voltage (B^2) in the ANOVA analysis, from Table 4, indicated that p-value less than 0.05 reflected the significance of the quadratic term value. The main effect graph indicated that the maximum microhardness of 238.29 HV was obtained at 19.25 V before it started to drop. Bodude and Momohjimoh [17] also studied the effect of welding voltage with constant current on microhardness value. As the voltage increased from 100 V to 220 V at 100 A, the microhardness value decreased. An increase in travel speed from 2 RPM to 4 RPM led to subtle change of weldment microhardness from 220 HV to only 223 HV. However, the quadratic term of travel speed (C^2) in the ANOVA analysis, from Table 4, indicated that p-value less than 0.05 reflected the significance of the quadratic term value. This was also indicated by the main effect graph where the highest microhardness value, at 3 RPM was 238.29 HV before it started to drop. Similar result was found by Kenchireddy et al. [18] as they increased the travel speed, the microhardness increased and started to drop at certain speed.

Table 4. ANOVA analysis of the quadratic model for response hardness

Source	Sum of Squares	Df	Mean Square	F Value	p-value Prob > F	
Model	8337.69	10	833.77	26.34	0.0004	significant
A-Arc Current	1582.20	1	1582.20	49.98	0.0004	
B-Arc Voltage	2.91	1	2.91	0.09	0.7721	
C-Travel Speed	18.07	1	18.07	0.57	0.4785	
AB	759.21	1	759.21	23.98	0.0027	
AC	430.40	1	430.40	13.60	0.0102	
BC	1305.15	1	1305.15	41.23	0.0007	
A^2	86.56	1	86.56	2.73	0.1493	
B^2	2577.08	1	2577.08	81.41	0.0001	
C^2	1189.11	1	1189.11	37.57	0.0009	
AB^2	1186.97	1	1186.97	37.50	0.0009	
Residual	189.92	6	31.65			
Lack of Fit	15.07	2	7.54	0.17	0.8476	not significant
Pure Error	174.85	4	43.71			
Cor Total	8527.61	16				
				R^2	0.9777	
				Adj. R^2	0.9406	
				Pred. R^2	0.8973	
				Adeq. R^2	15.940	

Interaction between arc current and arc voltage (AB) on microhardness was one of the main factors influencing microhardness value, with 0.0027 p-value. The interaction between arc current and arc voltage with respect to the weldment hardness showed that at low voltage, 17.0 V, the variation in arc current resulted in decreased of weldment microhardness value, from 217.82 HV to 197.93 HV. However, at high voltage, 21.5 V, variation in arc current significantly changed in weldment microhardness. It increased from 191.47 HV to 227.97 HV. This indicated strong interaction between two parameters. The ANOVA analysis revealed that the interaction between arc current and travel speed (AC) influenced the microhardness value. The interaction between arc current and travel speed with respect to the weldment microhardness showed that at low travel speed, 2 RPM, the variation in arc current resulted in minimal changes of weldment microhardness value, from 225.07 HV to 206.04 HV. However, at high travel speed, 4 RPM, variation in arc current significantly changed in weldment microhardness. It decreased from 248.82 HV to 188.30 HV. This indicated that arc current significantly affected the weldment microhardness with interaction of travel speed.

4. Conclusion

This study made a significant contribution to the field of knowledge related to TOW of dissimilar material of SS304 and BS1387 using GMAW as heat source. This study has found that the parameters used were able to produce good quality weldment on dissimilar material of SS304 and BS1387. The quality of the weldment was visually observed using dye penetrant method and the result shows no major defect on the weldment. However, there were a few flaws at start-end point on the weldment due to lack of fusion. Still, the other parts of weldment had no defect and were used for specimen preparation. The results of RSM indicated that the welding parameters did affect the mechanical properties of weldment. The tensile strength increased when the arc current and arc voltage increased and travel speed decreased. Similarly, the microhardness values also increased when the arc current and arc voltage increased and travel speed decreased.

Acknowledgement
Authors would like to thank the Faculty of Manufacturing Engineering, Universiti Teknikal Malaysia Melaka for providing facilities for this research to be conducted successfully. This research has been funded by the Ministry of Higher Education of Malaysia through a research grant with the reference number of FRGS/1/2015/TK03/FKP/02/F00280.

References
[1] Vandewynckéle A, Vaamonde E, Fontán M, Herwig P, Mascioletti A 2013 *Physics Procedia* **41** 144 – 152
[2] Chan S.M, Chan L C, Lee T C 2003 *J. Materials Processing Technology* **132** (1 - 3) 95 - 101
[3] Akbari D, Sattari-Far I 2009 *Int. J. Pressure Vessels and Piping*, 86(11), pp.769–776
[4] Hajiannia I, Shamanian M, Kasiri M 2013 *Materials and Design* **50** 566 - 573
[5] Joseph A, Rai S K, Jayakumar T, Murugan N, 2005 *Int. J. Pressure Vessels and Piping* **82**(9) 700 - 705
[6] Lee J, Han S, Kim K, Kim H, Lee U 2013 *Engineering Failure Analysis* **34** 300 - 307
[7] Zhao L, Liang J, Zhong Q, Yang C, Sun B, Du J 2014 *Advances in Engineering Software* **68** 70 - 79
[8] Bendick W, Gabriel J, Hahn B, Vandenberghe B 2007 *Int. J. Pressure Vessels and Piping* **84**(1-2) 13 - 20
[9] National Boiler Service 2016 *Boiler Generating Banks*
[10] Dwivedi S N, Sharan A 2003 *J. Materials Processing Technology* **141**(2) 155 - 162
[11] Siegel R 2004 *Inspection Trend* 17 - 20
[12] Talabi S I, Owolabi O.B, Adebisi JA, Yahaya T 2014 *Advances in Production Engineering and Management* **9**(4) 181 – 186
[13] Ampaiboon A, Lasunon O, Bubphachot B 2015 *The Scientific World Journal* **5**
[14] Ragu Nathan S, Balasubramanian V, Malarvizhi S, Rao A.G 2015 *Defence Technology* **11** 1 - 10
[15] Hussein N I S, Ayof M N, Nordin, S 2015 *Int. J. Materials, Mechanics and Manufacturing* **4**(1) 56 - 59
[16] Rohit J, A K. J 2014 *Int. J. Eng. Research and Application* **4**(6) 106 - 111.
[17] Bodude M A, Momohjimoh I 2015 *J. Materials Characterization and Engineering* **3** 142 - 153
[18] Kenchireddy K M, Jayadeva C T, Sreenivasan A 2014 *Global Journal of Engineering Science and Researches* **1**(3) 36 42

IOP Conf. Series: Materials Science and Engineering **238** (2017) 012016 doi:10.1088/1757-899X/238/1/012016

Variant analysis in coarse grain heat affected zone of low carbon steel

Mohammad Khairul Azhar Abdul Razab*, Sarizam Mamat, Muhammad Iqbal Ahmad and Nurul Syahida Mohd Nasir

Advanced Materials Research Cluster, Faculty of Bioengineering and Technology, Universiti Malaysia Kelantan Jeli Campus, Locked Bag No. 100, 17600 Jeli, Kelantan, Malaysiaof Mechanical Engineering, Universiti Teknologi Malaysia, 81310 Skudai, Johor Bahru, Malaysia

E-mail: azhar@umk.edu.my

Abstract. Coarse Grain Heat Affected Zone (CGHAZ) is a part of Heat Affected Zone (HAZ) that affected by heat during welding process. Application of different heat input dramatically varies CGHAZ microstructures without a noticeable changing in prior austenite grain size. The unique coarse microstructure and crack initiated at CGHAZ show there were possibility changes in crystallographic structure which may relate to the variant selection phenomenon. The aims of this study are to find the effects of heat input to the variant selection especially to the biggest grain at CGHAZ, hence correlate to the toughness and properties of the welded steel. The results show that heat input affected the variant selection at the biggest CGHAZ grain of low carbon steel. As heat input increase, grain area was increased and becomes coarser. Variant selection phenomenon present at low carbon steel due to the increase in grain diameter and high angle value. It was found that variant selections had occurred during the transformation.

1. Introduction

Carbon steel is most popular material used in large scale industries due to abundance source, low cost and variety mechanical properties. In order to meet those large scale industries such as shipping and automotive demand, the steel usually joined by welding technique. Usually in welding carbon steel, low carbon steel is the most preferred material based on weldability and machinable properties because most of their steel phase are ferrite and pearlite. There are many factors in produce secure weld joined, by manipulate welding heat input, defect and discontinuity can be reduce. According to Nasir (2016) [1], heat input is one of the dominant welding parameter should be considered in controlling the production of residual stress on welded material that lead to material defects. Moreover, the different applications in heat input results changes in microstructure, mechanical and physical steel properties especially at coarse grain heat affected zone (CGHAZ). Most crack initiate at CGHAZ due to the microstructural and property alteration that are influenced by the heat input [2].

The variant selection and formation may improve or cause defect to the steel quality. There are a lot of studies on the variant selection occurring at specific martensite or bainite grain/phase, but studies on variant formation/selection for other phases are still not well documented. Due to this issue, a study was conducted to unveil and correlate the effect of heat input to the variant and its formation at coarse grain heat affected zone (CGHAZ) of low carbon steel during phase transformation, which is known to have some effect on the toughness of the steel. In this study, low carbon steel undergoes Flux Cored Arc Welding (FCAW) process by using three designated heat input parameters which are low (0.99 kJ/mm), medium (1.22 kJ/mm) and high (2.25 kJ/mm) heat input. Electron backscatter diffraction (EBSD) was used to investigate variant analysis of low alloy carbon steels.

2. Experimental Procedure

Through this research, ABS Grade A steel plates were welded using filler wire. The chemical compositions for both materials were presented in Table 1 and 2, respectively. For this experiment, each steel plate was welded by three different heats inputs based on Muda (2015) [3] and welding procedure specification (WPS) of Sime Darby Engineering Sdn. Bhd. Table 3 shows the three parameters of heat inputs has been applied which are low (0.99 kJ/mm), medium (1.22 kJ/mm) and high (2.25 kJ/mm).

Table 1. ABS Grade A chemical composition

	Chemical Composition (%)								
Mill Certificate	C	Si	Mn	P	S	Cu	Ni	Cr	Mo
	0.17	0.14	0.49	0.018	0.005	0.02	0.02	0.03	0.01

Table 2. Filler wire chemical composition

Wire Designation	Wire Dimension (mm)	Chemical Composition (%)					Tensile Strength (N/mm^2)
		C	Si	Mn	P	S	
K-71T	1.2	0.04	0.55	1.25	0.015	0.011	580

Table 3. Welding parameters

Sample	Current (A)	Voltage (V)	Speed (mm/s)	Heat Input (kJ/mm)	Indicator
A	135	22	180	0.99	Low
B	166	22	180	1.22	Medium
C	196	23	120	2.25	High

2.1. Sample preparation

The samples were prepared using Flux Cored Arc Welding (FCAW) under the single pass welding method. The welding processes were carried out at the Faculty of Mechanical, Universiti Malaysia Pahang (UMP) using MIG machine model DM-500EF made by Ripson Machinery Sdn.Bhd., Malaysia and semi-automatic welding table as shown in Figure 1.

Figure 1. Equipment for semi-automatic welding machine.*(a)* Power source (down) and Flux Cored Arc Welding Machine (up); and *(b)* Automatic Welding Table.

IOP Conf. Series: Materials Science and Engineering **238** (2017) 012016 doi:10.1088/1757-899X/238/1/012016

2.2. *Electron backscatter diffraction (EBSD) analysis*

Electron backscatter diffraction (EBSD) is a microstructural-crystallographic characterization technique to study any crystalline or polycrystalline material. The technique involves understanding the structure, crystal orientation and phase of materials in the Field Emission Scanning Electron Microscope (FESEM). FESEM and EBSD were used to analyse grain size, grain misorientation angle, and variant analysis. EBSD became a popular method used to measure grain size because it can observe nano scale grain size [4]. Due to its flexibility, which can differentiate variants, EBSD became a famous method in variant selection studies. In this study, the samples were analysed using EBSD technique to determine its grain size, grain misorientation angle, and variant analysis. All the analysis was focused on the biggest grain at CGHAZ area for each heat input. The sample's single grain details such as grain size, misorientation angle, and variant analysis were analysed by using OIM analysis software.

2.2.1. *Grain size identification*

A single austenite grain is consisting of sub-grains or also called variants. Figure 2 shows that within single grain, there are sub-grains or variants with different crystallographic orientation. Variant can be considered as symmetry of crystals which is a distinct number of equivalent combinations of parallel planes and directions [5]. Sub-grains or variant is the 3D crystal within a sample which had different crystallographic orientation between its surrounding. The variants/grains size was identified by defining a critical misorientation angle and grain boundaries. In grain size EBSD analysis, the position of grain boundaries was determined by identifying changes in crystallographic orientation between neighboring grid points of greater than a defined minimum of 10°.

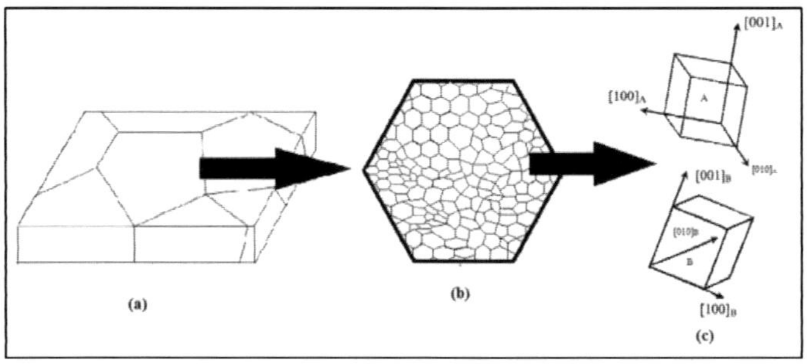

Figure 2. *(a)* Biggest grain at CGHAZ area;*(b)* sub grains or variants within single grain and*(c)* crystal orientation.

2.2.2. *Variant analysis*

Grain size and misorientation angle were analyzed using tools/application in OIM analysis software. Variant analysis also uses the same OIM analysis, but there were some additional steps that needs to be followed to obtain variant results. Variant formation in a single grain of targeted grain area was determined using inverse pole figure (IPF). The IPF of targeted grain area was then adjusted to the standard IPF as shown in Figure 3. The 24 standard variants of K-S orientation relationship were implemented to investigate the variant selection.

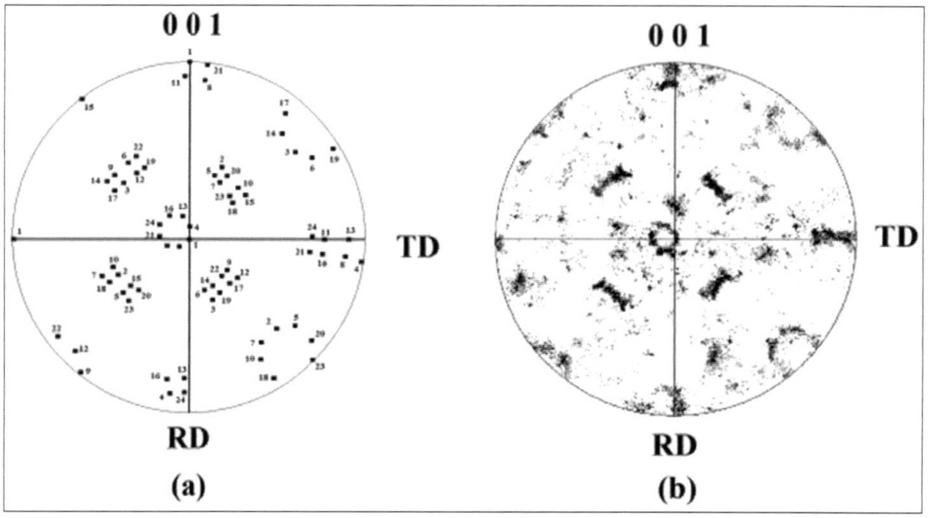

Figure 3. *(a)* Standard IPF - Configuration of <001>α pole based on K-S variants and*(b)* A 001 α pole figure of EBSD analysis data for low carbon steel under medium heat input which adjusted to (a) [6].

3. Results and Discussion

Variant was analyzed using electron backscatter diffraction (EBSD) to obtain quantitative microstructural crystallographic of steel. EBSD reveals grain size, grain boundary character, grain orientation, texture, and phase identity of steel. There are three criterias which can determine variant selection which were dual orientation, grain boundary plane orientation and growth selection [7]. Any of these criterias can be used to determine variant selection; however, variant selection would be more influenced by grain boundary plane orientation. This study focused on grain size and misorientation which are related to the grain boundary plane orientation.

3.1. Grain size

Figure 4 shows the summarized data for diameter for smallest, biggest and mode sub-grain under low, medium and high heat input of low carbon steel. The diameter for the smallest, biggest and mode grain for low carbon steel increased as the heat input was increased. Even though the grain mode diameter kept increasing, the percentage (quantity) of the mode decreased as heat input was decreased which were 18.23% for low heat input, 17.15% for medium heat input and 16.91% for high heat input. The same situation happened to the biggest grain; the grain percentage decreased as the heat input increased. From figures 4, it can be seen that majority grain diameter (mode) came from the small size diameter. These showed that variant selection had occurred, because certain grain (biggest grain) was preferable than others, the grain had the chance to transform or express more and made the grain bigger than other grains. Due to time and space constraint, some of undesired grains managed to transform and some other grains couldn't transform. This case caused the undesired grain to transform into different small grain size and some of the grain vanished from the CGHAZ grain.

IOP Conf. Series: Materials Science and Engineering **238** (2017) 012016 doi:10.1088/1757-899X/238/1/012016

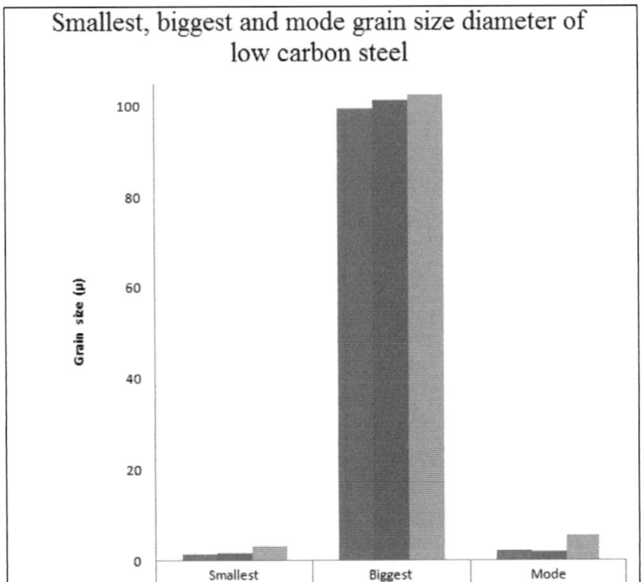

	Smallest	Biggest	Mode
Low heat input	1.33088	99.5601	2.09603
Medium heat input	1.62219	101.134	2.01635
High heat input	3.25245	102.598	5.60898

Figure 4Error! No text of specified style in document.. Diameter for smallest, biggest and mode sub-grain size of low carbon steel at low, medium and high heat input.

3.2. Misorientation

Misorientation is the transformation required to transform tensor quantities such as vectors, stress, and strain from one set of crystal axes to the other set. Misorientation happen when variants transform; due to time and space constraint, the variants keep pushing each other and cause variant overlaps with each other and make the variants have different orientation. This phenomenon causes the disturbance to its atomic structure of the grain boundary and leads the grain boundary to be in groups of low and high misorientation angle. Researchers have different opinions in low or high angle value. Low angle boundary can be 5° - 15°, ~ 5°, <12° and also can be between 10° - 15°, while, angle larger than 10° - 15° can be classifies as high angle grain boundary [8 - 11]. During EBSD statistical analysis, any angle exceeding 2° is considered as a boundary. Due to that, any grain boundary between 2° to 15° is grouped into low angle grain boundary and other grain boundary exceeding 15° (>15°) is under high angle grain group. Low angle grain boundary is related to two-dimension dislocation network which tilt or twist grain boundary. While high angle grain boundary relates more on coincidence site lattice (CSL) which had influences to the variant selection phenomenon.

Figures 5 - 7 represent fraction of misorientation angle for low carbon steel under low, medium and high heat input. From the figures, all three CGHAZ grain of low carbon steel under low, medium and high heat input were dominated by grains/variants with low misorientation angle (<15°), while high misorientation angle (>15°) was the minority. However, by increasing the heat input, misorientation angle with <15° kept decreasing with increasing misorientation angle >15°.

IOP Conf. Series: Materials Science and Engineering **238** (2017) 012016 doi:10.1088/1757-899X/238/1/012016

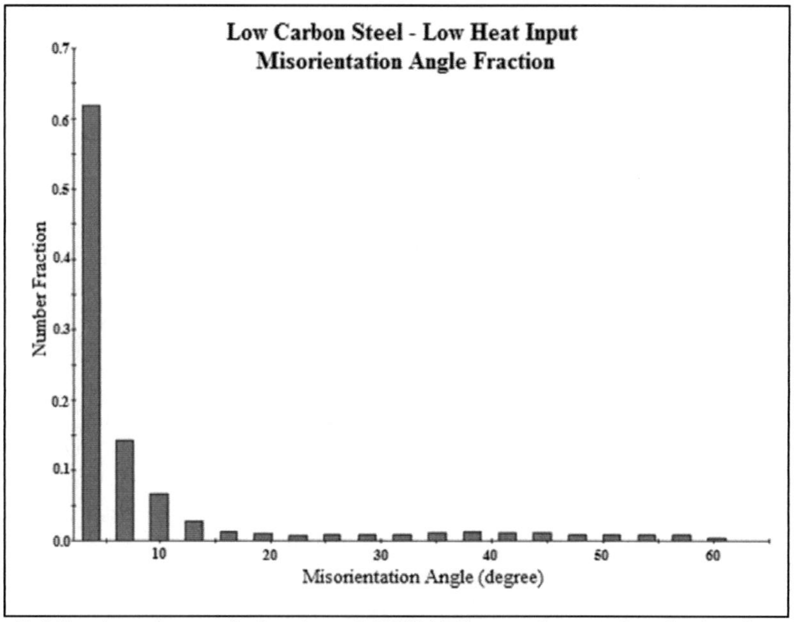

Figure 5. Misorientation angle fraction of CGHAZ low carbon steel under low heat input.

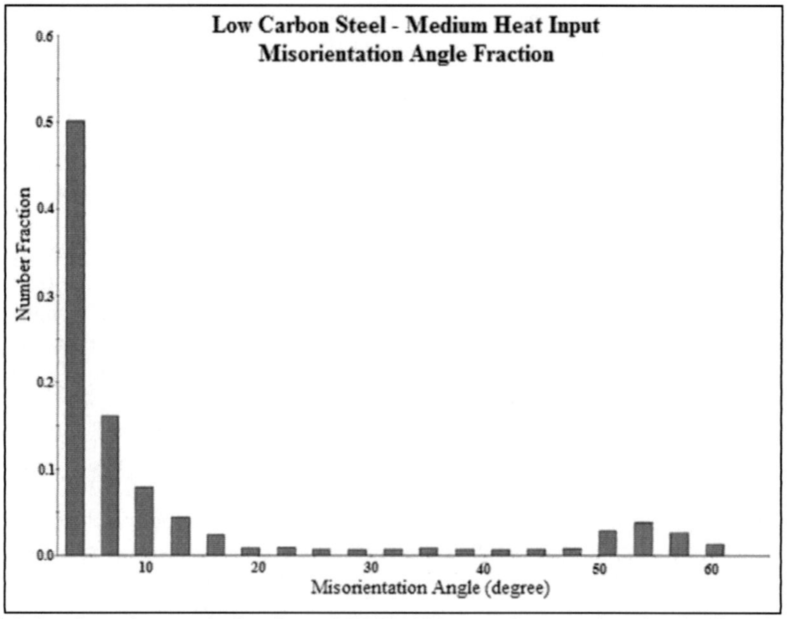

Figure 6. Misorientation angle fraction of CGHAZ low carbon steel under medium heat input.

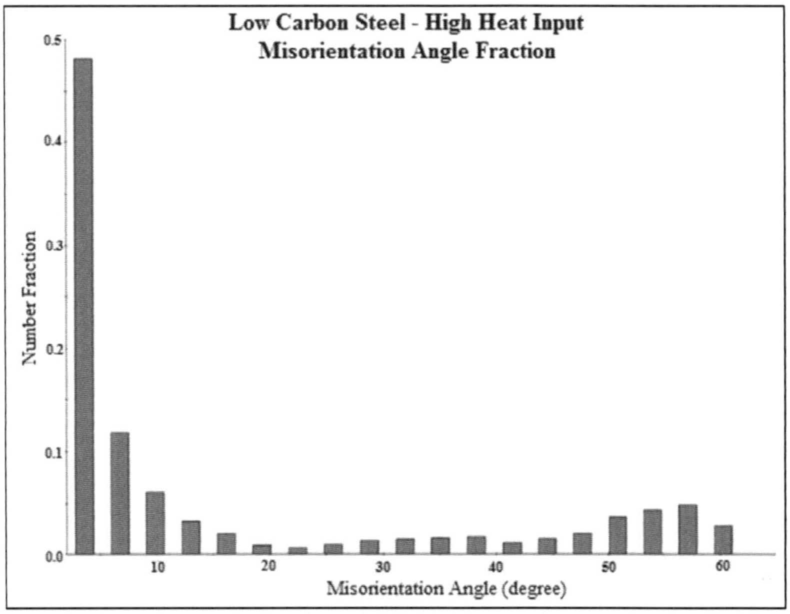

Figure 7. Misorientation angle fraction of CGHAZ low carbon steel under high heat input.

During transformation, grain growth occurred and filled all space within austenite grain boundary. When crystals or grains joined together, the crystal lattice couldn't complement each other's crystal and caused a grain boundary exists. Figure 8 shows atoms forming crystal grains and crystal grain boundary. The grain boundary of crystalline solids was very thin and might have one or two atoms. The grain boundary would have either low or high grain boundary, due to the selection that occurred during the transformation.

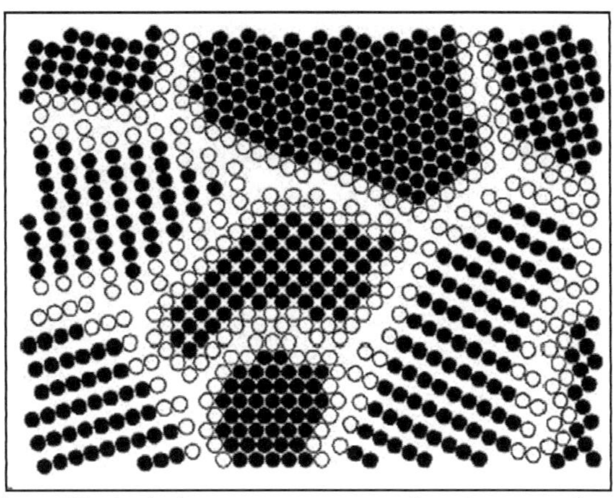

Figure 8. Grain boundary: Black circles represent atoms forming crystal grains, and white circles represent atoms located at crystal grain boundaries [12].

3.3. Variant selection

Variant selection is a phenomenon when certain variant within austenite grain boundary express more than other variants during the phase transformation. Single austenite consists of 24 variants; when variant selection occurs during the transformation there will be less than 24 variants present within the austenite grain boundary. Within the austenite grain boundary, the preferable and selected variant will

IOP Conf. Series: Materials Science and Engineering **238** (2017) 012016 doi:10.1088/1757-899X/238/1/012016

be in bigger size and the unpopular variant transform into smaller size and some of them vanish due to time and space constraint.

Figures 9 – 11 show the CGHAZ Inverse Pole Figure (IPF) images of variant distribution of low carbon steel under low, medium and high heat input. The IPF images show different color and each color represent different variant however the variant with maroon color was the biggest variant that occurred within the CGHAZ grain area. The preferable or selected variant for low carbon steel were V16, V15 and V1 for low, medium and high heat input, respectively. The biggest variant presence indicated the occurrence of the variant selection. Phase transformation without variant selection would have the number of variants the same as the number of variants before the transformation occurred. Moreover, the variant for phase transformation without variant selection mostly were in the same size. However, for phase transformation with variant selection, the selected or preferable variant would grow or express more during the phase transformation which led other unpreferable variants to have less time and space to grow and this caused some variant to vanish. In contrast, the phase transformation with variant selection would have less variant number than the original variant number before transformation occurred. Based on the Figures 9 – 11, the variant size kept decreasing with the increase of heat input. This shows that heat input had slight effect on variant sizes.

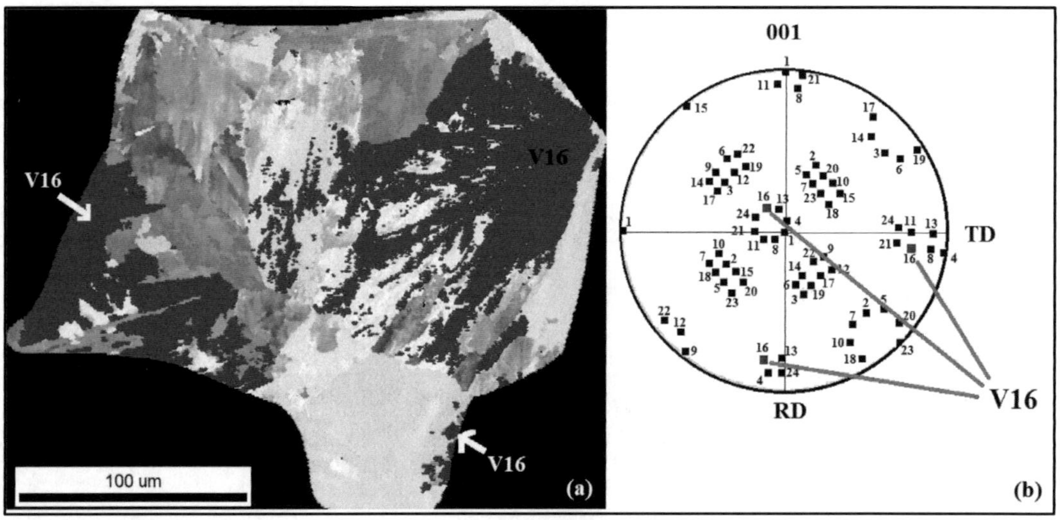

Figure 9. CGHAZ low carbon steel under low heat input. *(a)* IPF variant distribution and *(b)* 001 α pole figure of V16 location.

IOP Conf. Series: Materials Science and Engineering **238** (2017) 012016 doi:10.1088/1757-899X/238/1/012016

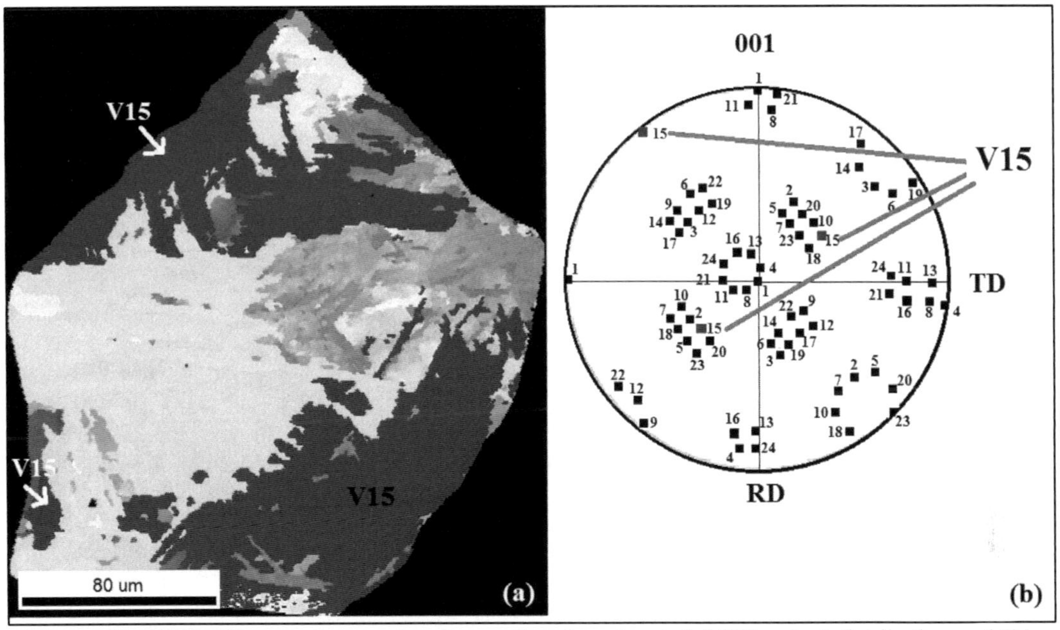

Figure 10. CGHAZ low carbon steel under medium heat input: *(a)* IPF variant distribution and *(b)* 001 α pole figure of V15 location.

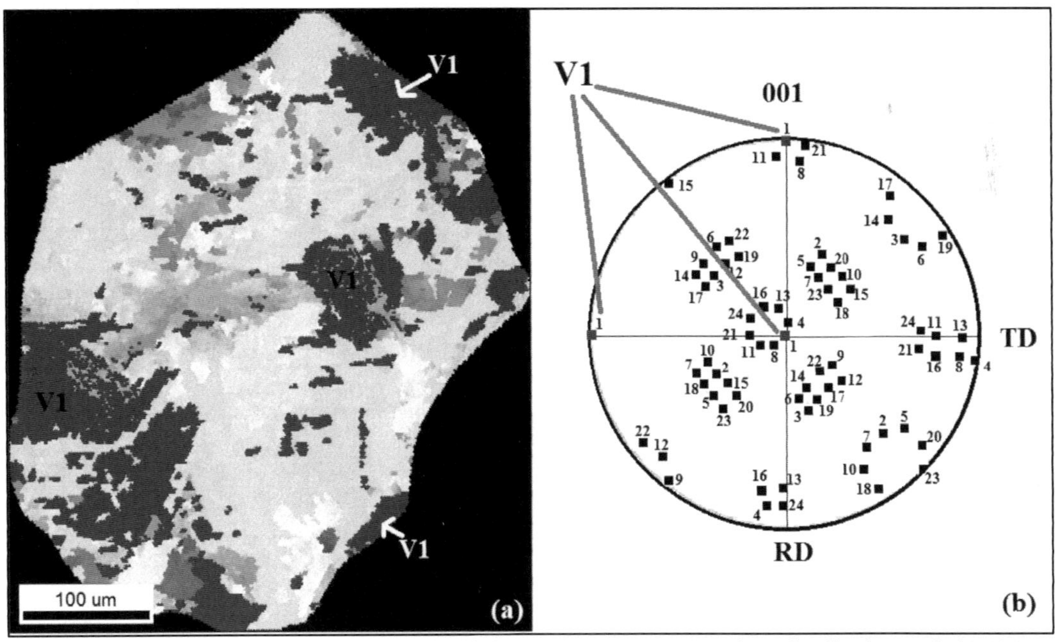

Figure 11. CGHAZ low carbon steel under high heat input: *(a)* IPF variant distribution and *(b)* 001 α pole figure of V1 location.

4. Conclusion

It was found that variant selections had occurred during the transformation. Based on significant data on variant grain diameter and high angle, it showed that heat input did affect variant selection at CGHAZ of low carbon steel. The result showed that by increasing the heat input, CGHAZ of low carbon steel hardness had decreased. This research proved that heat input affected variant selection and mechanical properties at CGHAZ of low carbon steel

Acknowledgement
Authors wishing to acknowledge technical assistance of Faculty of Mechanical, Universiti Malaysia Pahang (UMP) for providing the instruments and welding machines during the research works. We also acknowledge the support of the Short Term Research Grants (STRG) Scheme awarded by Universiti Malaysia Kelantan (UMK) under Grant No. R/SGJP/A08.00/00896A/001/2014/000160 and R/SGJP/A08.00/00451A/001/2015/000241.

References
[1] Nasir, N. S. M., Razab, M. K. A. A., Mamat, S., & Iqbal, M. 2016. Review on Welding Residual Stress. ARPN Journal of Engineering and Applied Sciences, 11, 6166 – 6175.
[2] Nasir, N. S. M., Razab, M. K. A. A., Ahmad, M. I., & Mamat, S. 2017. Influence of Heat Input on Carbon Steel Microstructure. ARPN Journal of Engineering and Applied Sciences, 12, 2689 – 2697.
[3] Muda, W. S. H. W., Nasir, N. S. M., Mamat, S., & Jamian, S. 2015. Effect of welding heat input on microstructure and mechanical properties at coarse grain heat affected zone of ABS Grade A Steel. ARPN Journal of Engineering and Applied Sciences, 10, 9487 – 9595.
[4] Oxford-Instruments. Grain size characterisation of a steel sample using the AZtecHKL EBSD system: Oxford Instruments; 2016 [11/10/2016]. Available from: www.oxford-instruments.com/ebsd.
[5] Youliang, H. E. Grain-scale characterization of fcc/bcc correspondence relations and variant selection. Montreal, Canada: McGill University; 2005.
[6] Sarizam, M., & Komizo, Y. 2014. Effects of Holding Temperature on Bainite Transformation in Cr-Mo Steel. Journal of Mechanical Engineering and Sciences, 7, 1103-1114.
[7] Kim, D. W. Variant selection of Allotriomorphic Ferrite in Steels. Pohang, Korea: Pohang University of Science and Technology; 2011.
[8] Lloyd, G. E., Farmer, A. B., & Mainprice, D 1997. Misorientation analysis and the formation and orientation of subgrain and grain boundaries. Tectonophysics, 279, 55-78.
[9] Saylor, D. M., Morawiec, A., Cherry, K. W., Rogan, F. H., Rohrer, G. S., Mahadevan, S., & Casasent, D 2002. Crystallographic distribution of low angle grain boundary planes in magnesium oxide. Materials Science Forum, 408, 1705-1710.
[10] Nucci, J. A., Keller, R. R., Field, D. P., & Shacham-Diamand, Y 1997. Grain boundary misorientation angles and stress-induced voiding in oxide passivated copper interconnects. Applied Physics Letters, 70, 1242-1244.
[11] Kamp, A. Investigation of Titanium a Plates by EBSD Analysis. Delft, Netherlands: Delft University of Technology; 2007.
[12] Japan Atomic Energy Research Institute. Persistent Quest Research Activities 2002 - 6.3 Radiation Resistance Properties of Nanocrystalline Materials 2002 [updated 01/05/2017]. Available from: http://jolisfukyu.tokai-sc.jaea.go.jp/fukyu/tayu/ACT02E/06/0603.htm.

IOP Conf. Series: Materials Science and Engineering **238** (2017) 012017 doi:10.1088/1757-899X/238/1/012017

Vibration analysis of resistance spot welding joint for dissimilar plate structure (mild steel 1010 and stainless steel 304

M.S.M.Sani[1,2], N.A.Nazri[1], D.A.J.Alawi[1]

[1]*Advanced Structural Integrity and Vibration Research (ASIVR), Faculty of Mechanical Engineering, Universiti Malaysia Pahang,*
26600 Pekan, Pahang, Malaysia
[2]*Automotive Engineering Centre (AEC), Universiti Malaysia Pahang,*
26600 Pekan, Pahang, Malaysia.

Email: mshahrir@ump.edu.my

Abstract: Resistance spot welding (RSW) is a proficient joining method commonly used for sheet metal joining and become one of the oldest spot welding processes use in industry especially in the automotive. RSW involves the application of heat and pressure without neglecting time taken when joining two or more metal sheets at a localized area which is claimed as the most efficient welding process in metal fabrication. The purpose of this project is to perform model updating of RSW plate structure between mild steel 1010 and stainless steel 304. In order to do the updating, normal mode finite element analysis (FEA) and experimental modal analysis (EMA) have been carried out. Result shows that the discrepancies of natural frequency between FEA and EMA are below than 10 %. Sensitivity model updating is evaluated in order to make sure which parameters are influences in this structural dynamic modification. Young's modulus and density both materials are indicate significant parameters to do model updating. As a conclusion, after perform model updating, total average error of dissimilar RSW plate is improved significantly.

1. Introduction

Welding is the process of joining metals with the presence of heat, with or without the application of pressure and filler metal [1]. A typical automotive structure could contain thousands of spot weld joints that contribute significantly to the vehicle's structural stiffness and dynamic characteristics [2]. Spot weld falls under fusion welding process by applying method of melting the materials for joining purpose. RSW has excellent paybacks such as low cost, high speed and suitability for automation which makes it an attractive choice for auto-body assemblies, truck cabins, rail vehicles and home appliances [3]. Joining of dissimilar materials is of risky as the process necessitates a proper matches concerning the chemical, physical, and mechanical properties between the materials placed together [4]. In order to achieve a strong connection between mild steel and stainless steel, the right welding parameter should be selected [5]. Joining dissimilar materials has recently become very popular in industries because of the advantages associated with the weld joint [6]. The main target of combining materials are to drive for more optimal, lightweight and high performance structure in order to gain product performance needed besides it is widely employed for different applications requiring certain special combination of properties as well as to save cost [7] [8].

Modal analysis is the process of characterizing the dynamic properties of an elastic structure by identifying its modes of vibration. It is a kind of exact forecast of structural dynamic features such as the natural frequencies and damping ratios in order to vibration control and evaluating their performance in a dynamic situation is a substantial phase in the design of structures which means each mode has a specific natural frequency and damping factor which can be identified from practically any point on the structure [9]. EMA focuses on impact hammer which is a method of testing that allows users to calculate the natural frequencies (modes), damping ratio and mode shapes of the test structure. In reality, impact hammer test is a simple method to implement but difficult to obtain consistent results as the force applied while hammering cannot be fixed throughout the experiment. In additions, this technique requires little equipment and can provide shorter measurement times by applying impulse. The top frequently used numerical method is finite element (FE) method which is utilized to simulate the behavior of real systems. The purpose of numerical modeling can be categorized into three different main parts which are analysis, prediction and design [10]. FEA simulation is a computerized procedure for the analysis of structures.

Finite element model updating, as described before in this thesis, is a procedure to minimize the differences between analytical and experimental results. It appears that it is very difficult to discard the errors and inaccuracies in finite element model when the structure in study is a complex structure. Several methods of structural model updating have been proposed and the topic is still under active study in various areas. Most of these studies centred on approaches such as the optimal matrix updating [11] and sensitivity-based parameter estimation [12, 13]. Meanwhile, there are many other approaches of model updating procedure that are being researched by numerous scholars such as Bayesian model updating [14] stochastic [15] and perturbation method [16]. All these techniques have been extensively developed, applied, studied and commented by many researchers and notable figure in model updating field [13, 17, 18].

Approximation subject to the simple first-order Taylor's series of expansion is used in model updating optimization in order to change the vector λ of eigenvalues based on the vector θ of structure updating parameters as stated in Equation 1. below:

$$\lambda_{i+1} = \lambda_i + [S_i](\delta\theta) \qquad 1.0$$

Where $[S_i]$ is a sensitivity matrix of ith iteration, which signify the rate of change of the structural eigenvalues λ_i with respect to changes in $\delta\theta$. The expression for the eigenvalue sensitivity S_i can be stated as Equation 2.0 as follows:

$$S_i = \frac{\delta\lambda_i}{\delta\theta} = \phi_i^T \left[\frac{\delta K}{\delta\theta} - \lambda_i \frac{\delta M}{\delta\theta}\right] \phi_i \qquad 2.0$$

The objective function, J, for the prediction error is defined as Equation **Error! No text of specified style in document.**.1:

$$J = \sum_{i=1}^{n} W \left(\frac{\omega_i^e}{\omega_i^a} - 1\right)^2$$

Error! No text of specified style in document..1

IOP Conf. Series: Materials Science and Engineering **238** (2017) 012017 doi:10.1088/1757-899X/238/1/012017

where ω_i^e and ω_i^a are the experimental and finite element natural frequencies respectively and W is the weighing factor for each mode.

In this paper, the main goal is to perform model updating based on sensitivity analysis of RSW dissimilar plate between two different types of materials; mild steel 1010 and stainless steel 304. This model updating is shown that modal properties can be improved significantly. Updating is completed when the error function minimized based on first five measured frequencies.

2. Modelling Plate Structure

2.1. Cutting materials

Mild steel and stainless steel are cut into specified dimension as depicted in Figure 1 by using shearing bench saw as shown in Figure 2 respectively below.

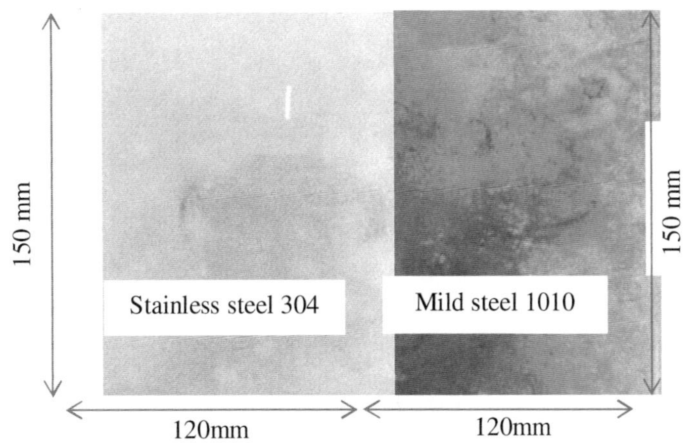

Figure 1. Stainless steel and mild steel

Figure 2. Shearing bench saw

2.2. Resistance Spot Welding Process

The history of welding initiate in the early twentieth century in which the localized heat source obtained from an electric arc that able to melts steel and coalesce two components together along a bond line to form a joint [19]. Resistance spot welding (RSW) is widely used for producing a typical BIW, that also being considered as dominant and important high speed process in automotive assembly which is normally made of thin metal sheets that are connected together by thousands of spot welds [20] .Welded structure strongly affected the global behavior for the whole structure and thus, it

IOP Conf. Series: Materials Science and Engineering **238** (2017) 012017 doi:10.1088/1757-899X/238/1/012017

is crucial to understand the dynamics characteristics of the welded joints which can be achieved via computational and experimental work [21]. In this paper, welding two dissimilar materials is need in which generally become more challenging than welding similar ones in term of differences in the chemical, physical and mechanical properties of the base metals even though the demands is accelerating [22, 23]. Set up parameter is a vital part to conduct welding process [5]. This prerequisite contributes to the quality and mechanical behavior of RSW in which might influenced the durability and crashworthiness of vehicle as it plays important role in joining parts in automotive [24]. Table 1 below shows the parameter set up for RSW in order to weld mild steel and stainless steel using RSW SSW-2020ATT machine as demonstrates in Figure 3.

Table 1. Parameter set up for RSW

Parameter	Value
Current supplied	4 A
Pressure applied	35 Psi
Holding time	5 s
Diameter of electrode tips	1 cm

Figure 3. RSW SSW-2020ATT machine

The completion of RSW process formed a joined between dissimilar plate with 1cm in diameter for each spot as can be seen in Figure 4 below.

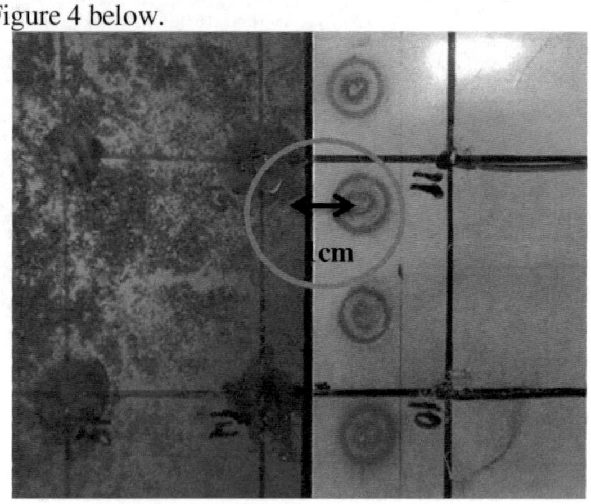

Figure 4. Diameter of spot weld

3. Finite Element Analysis

Finite element analysis is a very popular technique in structural dynamic analysis due to advances in numerical method and the availability of powerful computing facilities. The equation of motion can be showed dynamic properties of structure accompanying by easy geometric shape and standard physical characteristics. MSC Nastran Patran is a commercial finite element software package which having ability to model the specimen and utilize to generate its natural frequency and mode shapes using normal mode analysis (SOL103). The objective of numerical modeling can be classified into three different main parts which are analysis, prediction and design [25]. From analysis of dissimilar plate structure, modal parameters are gained and tabulated as in Table 2.

Table 2. Modal parameters using FEA

Mode	Natural Frequency (Hz)	Mode Shapes
1	114.23	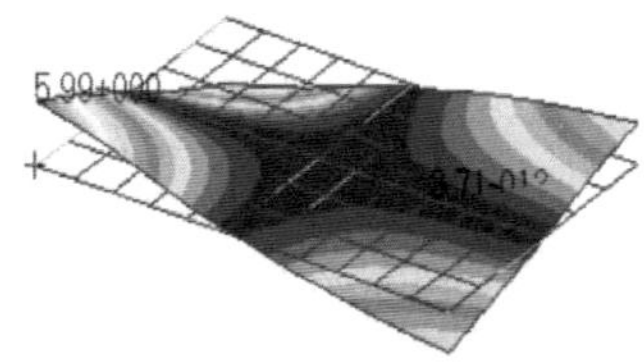
2	115.48	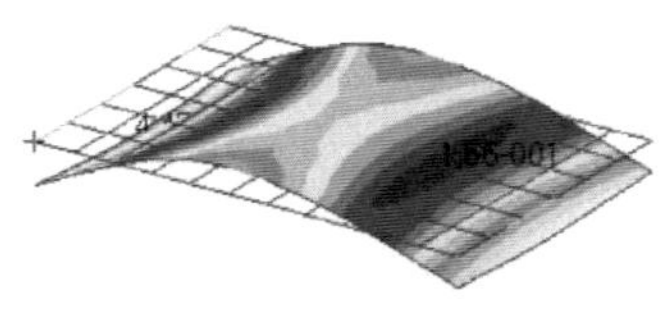
3	238.28	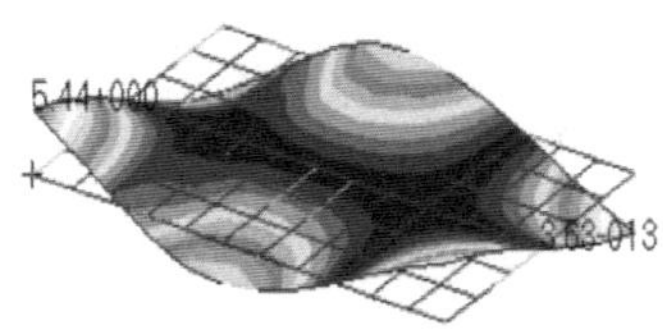
4	245.76	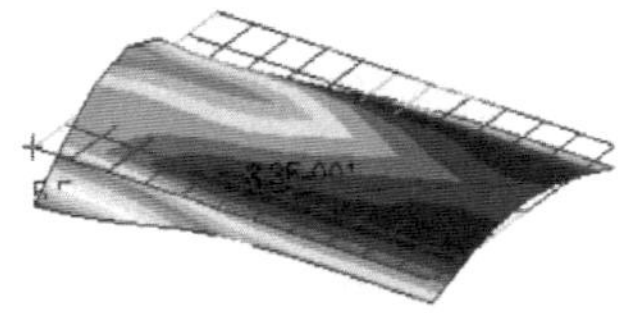
5	305.61	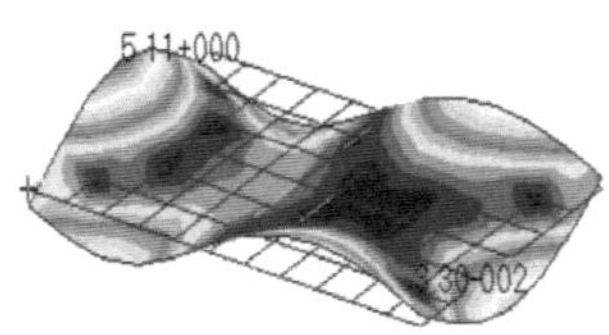

4. Experimental Modal Analysis

Experimental modal analysis or modal testing has grown steadily in popularity for the past several decades. Modal testing is defined as the study of dynamics characteristic of a mechanical structure. Even though modal analysis can also be done analytically through computational simulation, experimental modal analysis has the advantage of having modal characteristic defined from actual measurements and not from assumptions. An important property of modes or the measured frequency response functions from a modal testing can be used to describe the structure's dynamic properties.

IOP Conf. Series: Materials Science and Engineering **238** (2017) 012017 doi:10.1088/1757-899X/238/1/012017

The setup and the instrumentation used influence the experimental result in modal testing [26]. In order to gain accurate result of analysis, modal testing should be conducted in a free-free boundary condition. The experiment setup is illustrated in Figure 5.

Figure 5. Experiment setup for EMA

From the experiments, natural frequencies and mode shapes of dissimilar plates were acquired and tabulated in Table 3.

Table 3. Modal parameters using EMA

Mode	Natural frequencies (Hz)	Mode shape
1	119	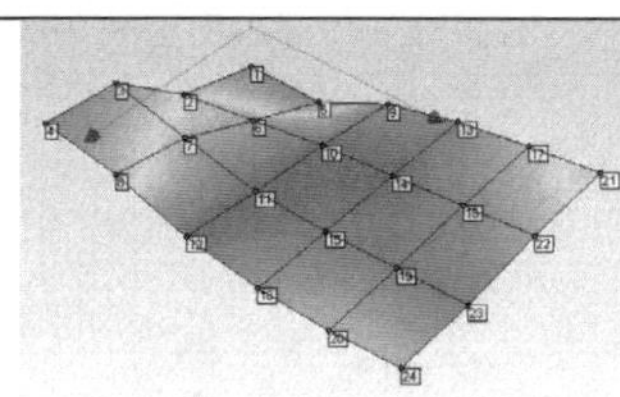
2	126	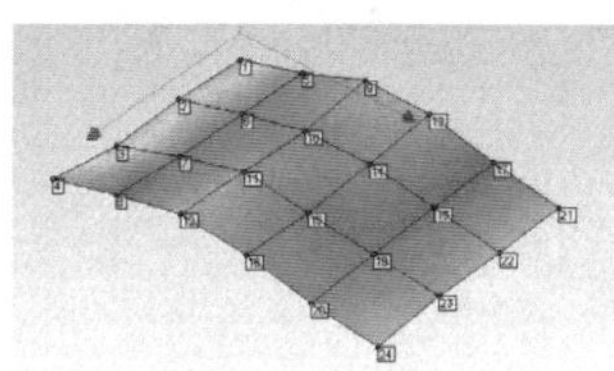
3	214	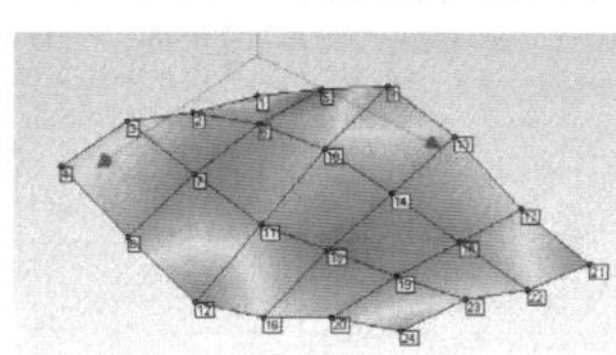

4	251	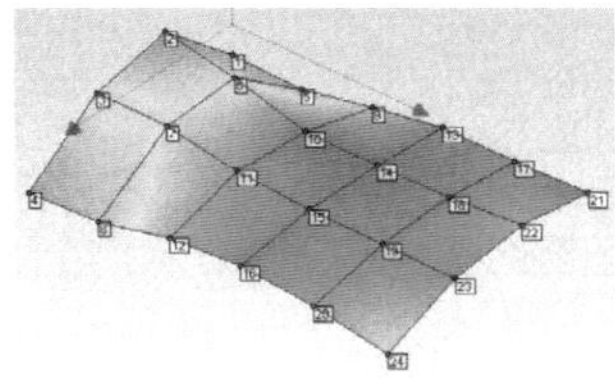
5	305	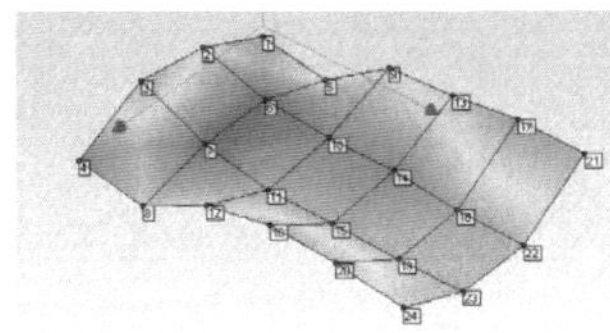

5. Model Updating

In model updating, there are two methods that can be implemented which is direct method and sensitivity method. In this paper, sensitivity method is chosen in which the most sensitive parameters will be selected for updating while less sensitive parameter is not taken into consideration for updating purpose. Eight parameters are having possibility to be updated undergo the sensitivity analysis using SOL200 in MSC Nastran software to select the suitable parameter to update. The sensitivities of the first five natural frequencies were tabulated in Table 4 and comparisons to be made in order to choose the most sensitive parameters.

Table 4. Sensitivity analysis for six parameters.

Output Type	**Young Modulus ($E_{mild\ steel}$)**	**Young Modulus ($E_{stainless\ steel}$)**	**Poisson Ratio ($v_{mild\ steel}$)**	**Poisson Ratio ($v_{stainless\ steel}$)**	**Density ($\rho_{mild\ steel}$)**	**Density ($\rho_{stainless\ steel}$)**
NF 1	27.00	31.23	-4.45	-5.82	-22.76	-35.65
NF 2	21.23	37.64	-0.52	1.42	-25.80	-33.29
NF 3	64.03	57.42	-0.10	-4.93	-61.42	-59.96
NF 4	50.76	74.93	5.26	8.68	-36.84	-89.61
NF 5	84.03	71.80	-11.64	-6.72	-83.20	-72.37

From all six parameters, it is clearly shown that poisson ratio, v is the less sensitive and will be not considered updating and optimization process. However, both dissimilar material of Young's Modulus, E and density, ρ show high level of updating sensitivity. These two need to be chosen and therefore, there are four parameters that is going to be updated which is Young's Modulus of mild steel, Young's Modulus of stainless steel, density of mild steel and density of stainless steel. Updating is performed by minimizing the error function and executed on the basis of the first five modes of measured frequencies. The number of updating parameters is kept to be less than number of modes updated in order to avoid ill-conditioning problem arising during updating procedure [16]. Table 5 displays the error between FEA and EMA after updating.

Table 5. Percentage of error between FEA and EMA after updating

Modes	**Experiment (Hz)**	**Initial FEA**		**Updated FEA**	
		FEA (Hz)	**Error (%)**	**FEA (Hz)**	**Error**
1	119	114.23	4.01	117.42	1.33
2	126	115.48	8.35	119.40	5.24

3	225	238.28	8.30	244.20	8.53
4	251	245.76	2.09	253.29	0.91
5	331	305.61	7.67	313.02	5.43
Total Average Error			6.08		4.29

From the table 5, it can be concluded that after updating process, the natural frequency improved and total average error become decreased. It is proved that the density and Young's Modulus which are considered as the sensitive parameter that significantly affect the modal properties of dissimilar plate structures. The changes of the initial and updated value of the updating parameters which consist of Young's modulus and density of both dissimilar materials are shown in Table 6.

Table 6. Updated value of parameters

Parameter	**Initial value**	**Updated Value**	**Deviation (%)**
Young's Modulus, $E_{mild\ steel}$ [GPa]	210	194	7.62
Young's Modulus, $E_{stainless\ steel}$[GPa]	200	210	5.00
Density, $\rho_{mild\ steel}$ (kg/m^3)	7870	7634	3.00
Density ($\rho_{stainless\ steel}$) (kg/m^3)	8000	7840	2.00

6. Conclusion

In summary, this study has deals with two approaches to obtain modal parameters which are FEA and EMA. From the result, discrepancies are appears and hence, model updating was executed in order to improve the correlation between measured counterparts. Before proceed to model updating, sensitivity analysis had been done to make sure the most sensitive parameters will be chosen for model updating. Results show that Young modulus and thickness are the most sensitive parameters for updating. After perform model updating, total error of the natural frequencies for dissimilar plates is reduced. As conclusion, model updating could be applied in order to reduce the discrepancies between numerical and experimental data, thus improving the dynamic structure to have better correlation with the actual structure.

References

[1] Raja R, Rajkumar M. 2015 A review on trouble shooting, testing and safety precautions procedure for welding process.

[2] Husain NA, Ouyang H. 2011 Detection of damage in welded structure using experimental modal data. Journal of Physics: Conference Series;305(1):012120.

[3] Akkaş N, İlhan E, Varol F, Aslanlar S. 2016 Welding time effect on mechanical properties in resistance spot welding of s235jr (Cu) steel sheets used in railway vehicles. Acta Physica Polonica A;129(4):541-3.

[4] Montemor M. 2016 Corrosion issues in joining lightweight materials: A review of the latest achievements. Physical Sciences Reviews;1(2).

[5] Engelmann C, Meier D, Olowinsky A, Kielwasse M. Metal meets Composite-Hybrid Joining for Automotive Applications. LIM; 2015.

[6] Ishak M, Idris SRA. 2014 Study of resistance spot welding between aisi 301 stainless steel and AISI 1020 carbon steel dissimilar alloys. Journal of Mechanical Engineering and Sciences (JMES);6:793-806.

[7] Martinsen K, Hu SJ, Carlson BE. 2015 Joining of dissimilar materials. CIRP Annals - Manufacturing Technology;64(2):679-99.

[8] Meshram SD, Paradkar AG, Reddy GM, Pandey S. 2017 Friction stir welding: An alternative to fusion welding for better stress corrosion cracking resistance of maraging steel. Journal of Manufacturing Processes;25:94-103.
[9] Adel F, Shokrollahi S, Jamal-Omidi M, Ahmadian H. 2017 A model updating method for hybrid composite/aluminum bolted joints using modal test data. Journal of Sound and Vibration;396:172-85.
[10] Zahari SN, Zakaria AAR, Sani MSM. 2015 A review on model updating of joint structure for dynamic analysis purpose.
[11] Berman A, Nagy EJ. 1983 Improvement of a Large Analytical Model Using Test Data. AIAA Journal;21(8):1168-73.
[12] Bakir PG, Reynders E, De Roeck G. 2007 Sensitivity-based finite element model updating using constrained optimization with a trust region algorithm. Journal of Sound and Vibration;305(1):211-25.
[13] Mottershead JE, Link M, Friswell MI. 2011 The sensitivity method in finite element model updating: A tutorial. Mechanical Systems and Signal Processing;25(7):2275-96.
[14] Mthembu L, Marwala T, Friswell MI, Adhikari S. 2011 Model selection in finite element model updating using the Bayesian evidence statistic. Mechanical Systems and Signal Processing;25(7):2399-412.
[15] Abu Husain N, Haddad Khodaparast H, Ouyang H. 2012 Parameter selection and stochastic model updating using perturbation methods with parameter weighting matrix assignment. Mechanical Systems and Signal Processing;32:135-52.
[16] Ouyang H, Mottershead JE, Haddad Khodaparast H, Abu Husain N. 2010 Application of the Perturbation Method With Parameter Weighting Matrix Assignments for Estimating Variability in a Set of Nominally Identical Welded Structures. (49194):95-104.
[17] Friswell MI, Mottershead JE, Ahmadian H. 2001 Finite–element model updating using experimental test data: parametrization and regularization. Philosophical Transactions of the Royal Society of London Series A: Mathematical, Physical and Engineering Sciences;359(1778):169-86.
[18] Mottershead JE, Friswell MI. 1993 Model Updating In Structural Dynamics: A Survey. Journal of Sound and Vibration;167(2):347-75.
[19] Moore P, Booth G. The Welding Engineer's Guide to Fracture and Fatigue. United Kingdom: Woodhead Publishing; 2015.
[20] Eisazadeh H, Hamedi M, Halvaee A. 2010 New parametric study of nugget size in resistance spot welding process using finite element method. Materials & Design;31(1):149-57.
[21] Zahari SN, Sani MSM, Ishak M, editors. Finite element modelling and updating of friction stir welding (FSW) joint for vibration analysis. MATEC Web of Conferences; 2017: EDP Sciences.
[22] Wu W, Hu S, Shen J. 2015 Microstructure, mechanical properties and corrosion behavior of laser welded dissimilar joints between ferritic stainless steel and carbon steel. Materials & Design (1980-2015);65:855-61.
[23] Kimura M, Suzuki K, Kusaka M, Kaizu K. 2017 Effect of friction welding condition on joining phenomena and mechanical properties of friction welded joint between 6063 aluminium alloy and AISI 304 stainless steel. Journal of Manufacturing Processes;26:178-87.
[24] Marashi P, Pouranvari M, Amirabdollahian S, Abedi A, Goodarzi M. 2008 Microstructure and failure behavior of dissimilar resistance spot welds between low carbon galvanized and austenitic stainless steels. Materials Science and Engineering: A;480(1–2):175-80.
[25] Zahari SN, Zakaria AAR, Sani MSM, Zaman I. 2015 A review on model updating of joint structure for dynamic analysis purpose.
[26] Ewins D. 2003 Modal testing: theory, practice and application (mechanical engineering research studies: engineering dynamics series).

IOP Conf. Series: Materials Science and Engineering **238** (2017) 012018 doi:10.1088/1757-899X/238/1/012018

Effect of tool offsetting on microstructure and mechanical properties dissimilar friction stir welded Mg-Al alloys

Amir Hossein Baghdadi *, Nor Fazilah Mohamad Selamat, Zainuddin Sajuri

*Department of Mechanical and Material Engineering, Faculty of Engineering and Built Environment, Universiti Kebangsaan Malaysia, 43600 Bangi, Selangor, Malaysia.

E-mail : baghdadi.amirhossein@gmail.com

Abstract. Automotive and aerospace industries are attempting to produce lightweight structure by using materials with low density such as aluminum and magnesium alloys to increase the fuel efficiency and consequently reduce the environmental pollution. It can be beneficial to join Mg to Al to acquire ideal performance in special applications. Friction stir welding (FSW) is solid state welding processes and relatively lower temperature of the process compared to fusion welding processes. This makes FSW a potential joining technique for joining of the dissimilar materials. In this study, Mg-Al butt joints were performed by FSW under different tool offset conditions, rotation rates (500-600 rpm) and traverse speeds (20 mm/min) with tool axis offset 1 mm shifted into AZ31B or Al6061 (T6), and without offset. During the welding process AZ31B was positioned at the advancing side (AS) and Al6061 (T6) was located at the retreating side (RS). Defect free AZ31B-Al6061 (T6) dissimilar metal FSW joints with good mechanical properties were obtained with the combination of intermediate rotation rate and low traverse speed pin is in the middle. When tool positioned in -1 mm or +1 mm offsetting, some defects were found in SZ of dissimilar FSWed joints such as cavity, tunnel, and crack. Furthermore, a thin layer of intermetallic compounds was observed in the stir zone at the interface between Mg-Al plates. The strength of the joint was influenced by FSW parameters. Good mechanical properties obtained with the combination of intermediate rotational speed of 600 rpm and low travelling speed of 20 mm/min by locating Mg on advancing side when pin is in the middle. Also, Joint efficiency of the welds prepared in the present study was between 29% and 68% for the different welding parameters.

1. Introduction

Nowadays, the automotive and aerospace industries are attempting to produce lightweight structure by using aluminum and magnesium alloys which have low density compare to other structure materials, to increase the fuel efficiency and consequently reduce the environmental pollution. Combination and assembly of these two alloys or with other metals through new design concept could results in better strength and performance. Thus, it can be beneficial to join Mg to Al in order to acquire ideal performance in special applications. However, the formation of intermetallic compound (IMC) in the interface of Al-Mg is unavoidable in the welding zone [1–6].

Until today, many researches have performed different joining methods to joint aluminum to magnesium alloys such as using laser weld bonding [7], TIG fusion welding [8], vacuum diffusion bonding [9], laser welding [10], etc. However, utilization of these methods for joining of Al-Mg alloys are limited because of lower strength of welded joint and particular process prerequisites. Recently, many researchers focused on friction stir welding to join aluminum to magnesium alloys [4,6,11].

Friction stir welding (FSW) is solid state welding processes invented by TWI in 1991 that can be performed to produce sound joint. FSW has lower temperature process due solid state nature compare to fusion welding such as MIG, TIG, and laser welding. So, FSW is believed that can be a potential method for welding of heterogeneous materials with reasonable mechanical properties. During FSW, high pressure accompanied by high shear strain can form metallurgical bands between bar metal surfaces which are positioned in close contact [11].

Many researches have been paid particular attention to tailoring the interfaces of dissimilar materials to enhance the mechanical properties. FSW is believed which has ability to make the complex nonlinear weld interface under certain welding conditions [11–13]. There are several reports about the joining of magnesium alloys to aluminum alloys by using FSW which some of the researches have mentioned about tool offsetting effect during the welding [6,11,12,14–21]. Venkateswaran and Reynolds [11] have reported which with 2 mm tool offset to Al6063, mechanical properties will decrease because of lower intermixing od Al and Mg in the interface and shorter zigzag line in the interface. Banglong Fu et al. [16] Believed that with tool offsetting about 0.3 mm to Mg side which has positioned in advancing side, sound Al–Mg dissimilar joints with welding efficiency of 70 % metal could be obtained. In other research performed by Firouzdor and Kou [21] have, they have reported mechanical properties of Al-Mg joint will be higher when the tool shifted 1.5 mm to AZ31 and Mg placed in advancing side by reducing the heat input and hence the detrimental formation of intermetallic compounds and liquid films. In another study which has been done by Firouzdor and Kou [19], they have mentioned that with tool positioned at the middle mechanical properties was slightly higher than tool had 1.5 mm offset to Mg side, in rotational speed of 1400 rpm when Al was in the advancing side. Dorbane et al. [22] showed with having aluminum on advancing side of the FSW of Al-Mg joints resulted in better welds with fewer defects in the stir zone and higher tensile strength when the pin was in the middle.

Therefore, the relationship between tool offsetting and welding properties for FSWed Mg-Al dissimilar metal need to be studied further. In this paper, Mg-Al butt joints were obtained by FSW under different tool offsetting and rotational speeds. The aim of this research was to evaluate the nature of dissimilar metal FSW of Al6061(T6)-AZ31B and effect of tool offsetting on joint strength.

2. Experimental Procedure

AZ31B magnesium alloy and commercial aluminum alloy 6061(T6) plates with dimensions of 75×50×4 mm used in this study as the base metal. The alloy compositions and mechanical properties of the materials used are mentioned in table 1. Prior to FSW, the surface oxides were removed by steel brush and then the surface was cleaned using ethanol. Figure 1 shows the joint configuration and welding parameters direction.

The welding direction is made perpendicular to rolling direction of AZ31B and Al6061 (T6) plates. The tool for FSW is a cylindrical threaded M5 pin made of a heat treated H13 tool steel with a hardness of 53-55 RC. Figure 2 shows the schematic view FSW tool pin with 3 degree tilled angle of tool axis. First, the tool was gradually pushed into the base metal sheet until the shoulder tip penetrated 0.2-0.3 mm to the base metal. During the welding process AZ31B was positioned at the advancing side (AS) and Al6061 (T6) was located at the retreating side (RS). Then, the tool started to be stirred along the joint. Tool rotation speed was varied from 500 to 600 rpm with tool axis offset 1 mm shifted into AZ31B, Al6061 (T6), and without offset. Figure 3 shows the tool position in different offsetting. +1 mm, 0 mm, and -1 tool position means tool shifted 1mm into AZ31B, tool in the middle, and tool shifted 1 mm into Al6061 (T6), respectively.

Table 1. Chemical compositions and mechanical properties of materials used

Material	Chemical composition (wt. %)						Yield stress (MPa)	Tensile stress (MPa)
	Al	Mg	Si	Mn	Cu	Zn		
Al6061(T6)	Bal.	0.8	0.4	0.03	0.18	-	258	293
AZ31B	2.5-3.5	Bal.	-	0.2-1.0	-	0.7-1.4	166	229

The workpieces were clamped down tight with four steel pads located 20 mm away from joint line. The fixture was used to minimize gap opening during the FSW as shown in figure 1. The sheets were held tightly against each other by tightening three bolts against the Al part in addition to being clamped tightly against the steel base by the hold-down clamps. Also, steel band with thickness of 3 mm was welded to the backing steel plate instead of using three bolts in AZ31B side. Microstructure observation was performed perpendicular to the welding direction. Then, samples were cold mounted, manually ground from #600 to #2000 and polished by using Al203 with the particle size of 1, 0.3, and 0.05 µm, respectively. An etching solution consisting of 10 ml acetic acid, 6 g picric acid, 10 ml distilled water, and 100 ml ethanol was used to reveal the microstructure of AZ31B. The second step was to etch sample with a solution consisting of 10 g NaOH in 50 mL distilled water for 45 seconds to reveal the phases in different colors. Microstructures of stir zone (SZ), thermo-mechanically affected zone (TMAZ) were observed by optical microscope (OM). Average grain size was measured based on the ASTM Standard E112-10. Tensile test was performed in ambient temperature (25 °C) with speed of 0.1 mm/min using specimen with length and width of 100 mm and 10 mm, respectively, which were cut perpendicular to welding direction.

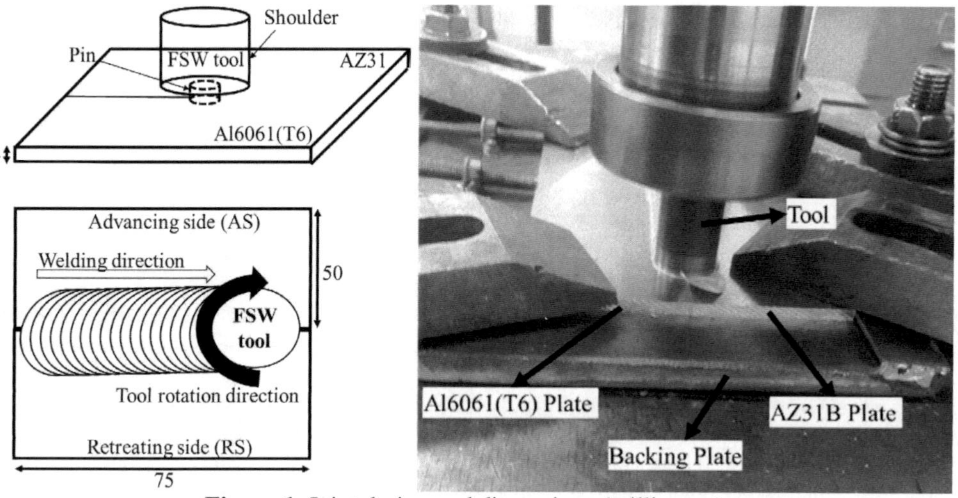

Figure 1. Joint design and dimensions (millimeter).

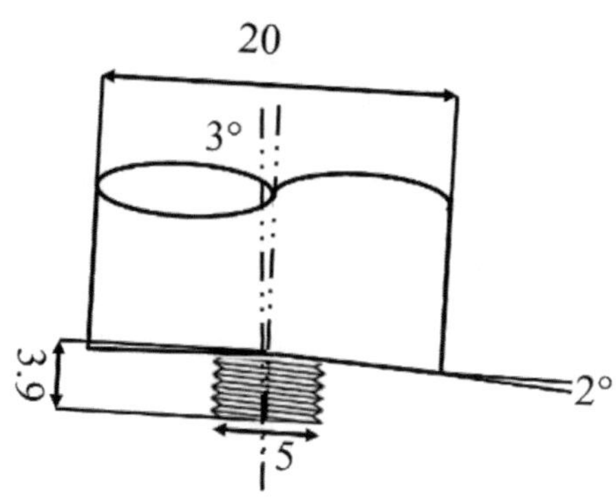

Figure 2. Schematic representation of welding tool (millimeter).

IOP Conf. Series: Materials Science and Engineering **238** (2017) 012018 doi:10.1088/1757-899X/238/1/012018

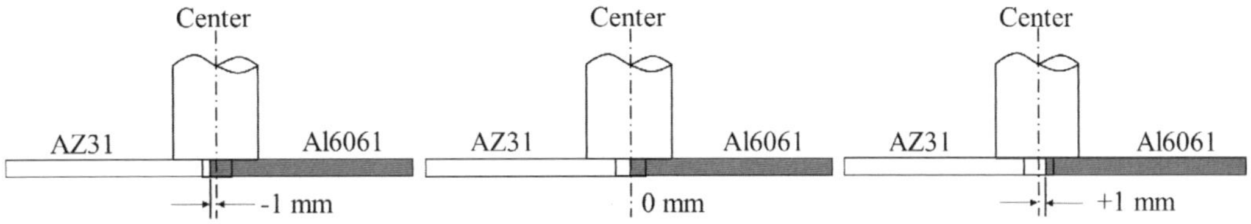

(a) -1 mm offset (b) 0 mm offset (In the middle) (c) +1 mm offset

Figure 3. Tool position at different offset.

3. Results and Discussion

3.1. Appearance and macrostructure

Figure 4 shows the typical cross section macrostructures of dissimilar friction stir welding (FSW) of AZ31B and Al 6061(T6) in rotation speed of 500 and 600 rpm and travelling speed of 20 mm/min with pin offsetting. The AZ31B was in advancing side and Al6-061(T6) was in the retreating side. For Mg-Al dissimilar joint when tool positioned in the middle, the weld appearance was smooth semicircular traces and less flash was formed and pushed out from the welding zone during the welding. With tool shifting into Al or Mg, the weld appearances were acceptable but the amount of flash formed were higher in compare to tool position in the middle. Moreover, the amount of the flash formed in -1 mm tool position was more than +1 mm tool position. Also, increasing the rotational speed from 500 rpm to 600 rpm was caused the increasing amount of flash expelled out from the welding.

In the macroscopic photo in figure 4, the left side is AZ31B and the right side is Al6061(T6). The boundary between the stir zone (SZ) and the thermo-mechnical affected zone (TMAZ) is clear and sharper on the Mg side. In the macrograph, there is zigzag line which shows the intermixing of Al and mg in the stir zone. When tool was in the middle of welding line, sufficient intermixing was observed in compare to tool shifted into Al or Mg sides. Moreover, with shifting of tool to Al side, the zigzag line was more irregular in compare to -1 mm tool position. Furthermore, the free defect joint was formed in when the tool was in the middle.

When tool was shifted into Al or Mg, the different welding defects which included of crack, cavities, and tunnel, were formed in the welding zone. Figure 5 displays the defects formed in rotational speed of 600 rpm and +1 mm tool position. The defects volume in +1 mm tool position was bigger than in the -1 mm tool position. The defects were made just in the 1 mm from the bottom side, and only in the one condition, crack was made in the middle of the thickness in the rotational speed 600 rpm and 20 mm/min of travelling speed as shown in Figure 5.

Rotational Speed	Tool offset (mm)		
	-1	In the middle (0)	+1
500 rpm	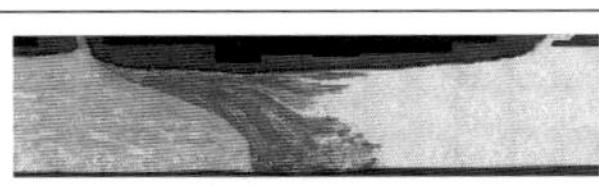	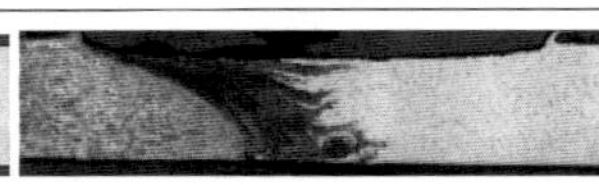	
600 rpm			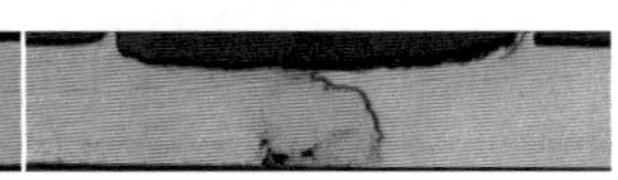

Figure 4. Cross section macrostructures of dissimilar FSW of AZ31B and Al 6061(T6) at travelling speed of 20 mm/min and rotational speed.

3.2. Microstructure

Figure 6 shows the transvers cross section macroscopic photo and microstructure of dissimilar friction stir welded Mg-Al sample with welding parameters of 600 rpm and 20 mm/min and tool in the middle. In the micrograph the left side is AZ31B and the right side is Al6061(T6). Figure 6(a) revealed the microstructure of AZ31B base metals on the advancing side. Microstructure of AZ31B revealed typical rolled microstructure of Mg alloys with inhomogeneous grains and numerous twins. The average grain size of the AZ31B was evaluated about 84 μm.

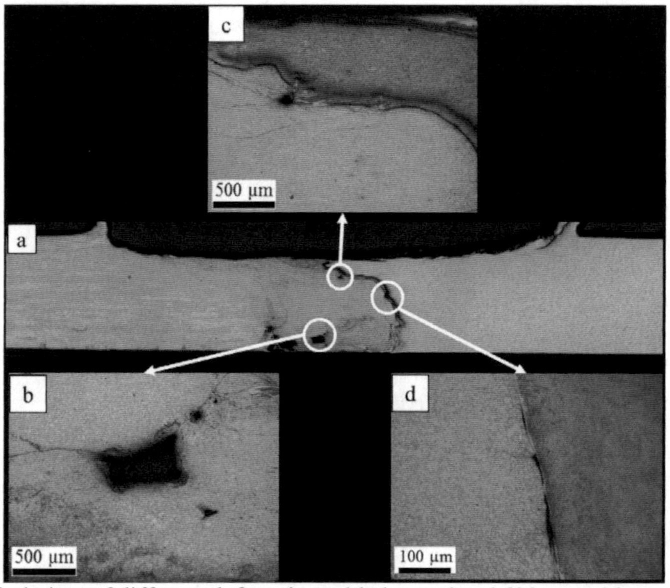

Figure 5. Magnified view of different defects in welded sample of 600 rpm, 20 mm/min, and +1 mm tool position: a) cross section, b) Tunnel volumetric defect, c) Cavity volumetric defect, d) Crack defect along the interface.

Microstructure of Mg-Al FSWed sample is divided in 3 zones which are heat effected zone (HAZ), thermomechanical effected zone (TMAZ), and nugget zone (NZ) or stir zone (SZ) as shown in figure 6b). Furthermore, Three distinct zones were defined in nugget zone of dissimilar friction stir welded Mg-Al samples; the shoulder affected zone (figure 6(c)) which was just below the shoulder during welding, the banded zone (figure 6(f)) which was in the upper and middle parts of the joints on AS, and severe intercalated zone (figure 6(g)) which was in the bottom which have been reported in recent researches [23] and involved lamellar-like shear bands. It also can be seen that intercalated structures formed no matter at the upper part of the joint or at the lower part of the joint are all sound without defects.

In the magnesium side, the TMAZ shows a larger number of twins and is distinguished with more elongated, bigger grains and partially equiaxed grains compared to the AZ31B base metal. Extreme plastic deformation and frictional heating during FSW lead to the generation of fine recrystallized grains in the SZ which was shown in figure 6(e). It will produce a finer grain structure, either decreasing the temperature or raising the strain rate.

Material flow in the shoulder affected zone was dominated by the plastic flow of material on AS with the direction of tool rotation, in this case, Mg penetrated into Al side under the mechanical stirring action of tool shoulder. Figure 6(d) shows the element of Al6061(T6) in SZ of AZ31B side which is a good evidence for intermixing of Al and Mg in SZ at Mg side. Furthermore, EDS mapping analysis was performed on the location of (I) as shown in the figure 7. It shows the main elements distribution of Al and Mg in that zone. Also, figure 6(h) shows a typical microstructure of dissimilar Mg-Al weld interface. The white line at the middle of interface clearly shows of intermetallic Compound (IMC) which have been formed in the interface of joint by mechanical stirring and frictional heat [6,16,20]. Constitutional liquation and solidification are caused to form intermetallic compound $Mg_{17}Al_{12}$ during FSW [13].

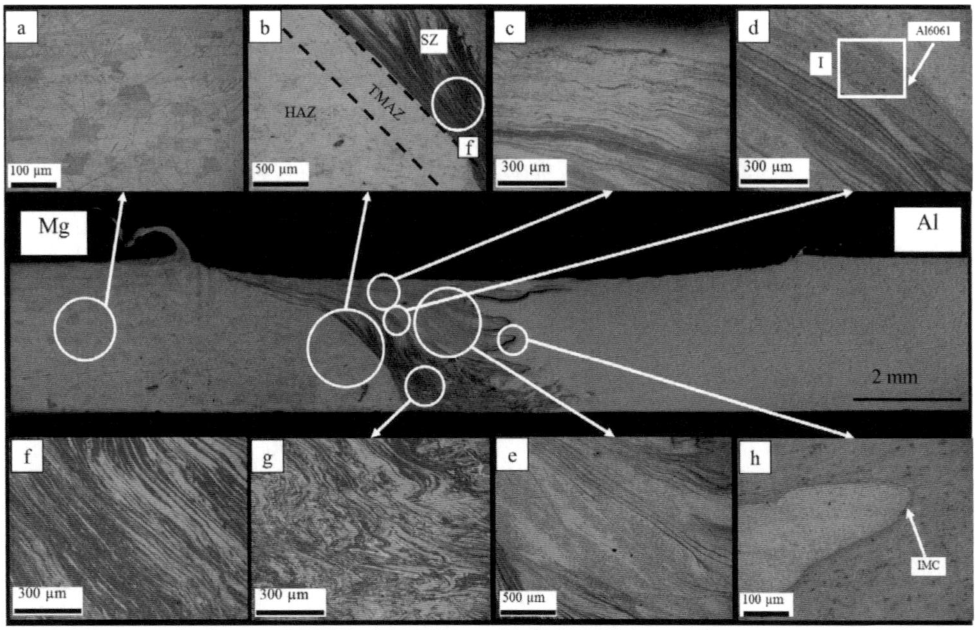

Figure 6. Micrograph photos of FSW of AZ31B to Al 6061(T6) in welding parameters of 600 rpm, 20 mm/min, and tool in the middle.

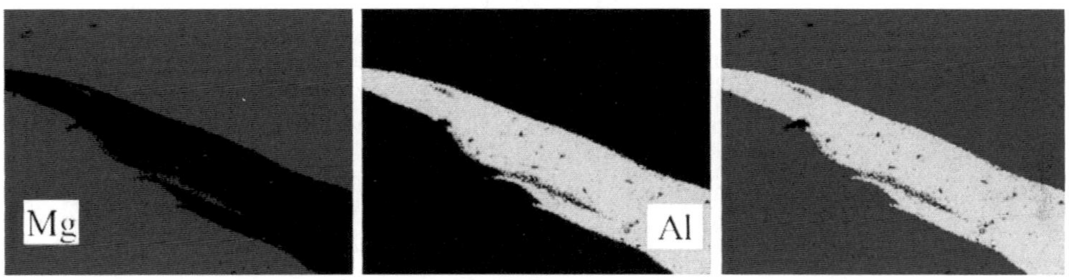

Figure 7. EDS mapping analysis of location (I) in figure 5(d)

3.3. Tensile properties

The transverse tensile properties of FSWed joints were obtained to evaluate the tensile strength, elongation, and welding efficiency under different welding parameters. The specimens display before tensile testing in Figure 8. In each condition, three specimens were tested and the average of three results are presented in table 2. As can be seen, the tensile strength of Mg-Al dissimilar FSW joints were lower than Al and Mg base metal which has mentioned in Table 2.

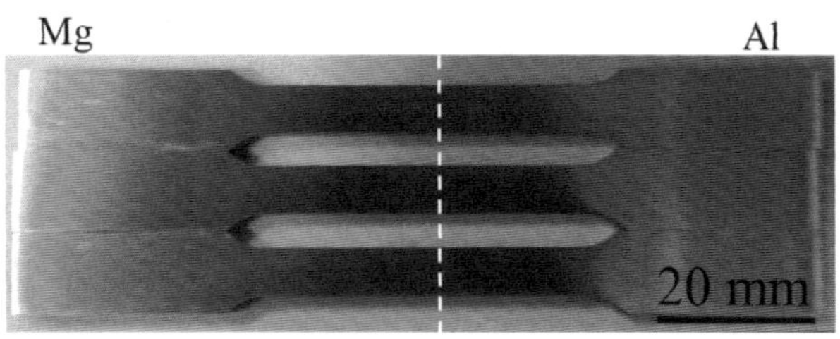

Figure 8. Tensile sample before test

The tensile strength of welded samples was varied by shifting of tool. When pin was in the middle, the highest tensile properties was achieved and tensile properties were decreased with shifting of tool into

Al6061(T6) or AZ31B. Also, the lowest tensile strength was obtained when tool was in +1 mm tool position. It was mentioned by recent research[16] where tool offset more than 0.3 mm/min, the tensile strength will decrease. It is due to lower heat input by shifting tool in Al6061(T6) or AZ31B and lower heat input is caused less mixing of Al and Mg in the stir zone in compare to tool in 0 mm tool position. Furthermore, distribution of intermetallic on the welding zone is effective on mechanical properties of Mg-Al joints[24] as shown in figure 5(d) which was caused a defect and reduced the tensile strength. When tool is at the middle, it is produced higher heat input and better mixing of Al and Mg in the welding zone. When tool shifted into AZ31B, mechanical properties of welded samples were increased by increasing of rotation speed, similar to -1 tool position [16].

Table 2. Transverse tensile properties of FSWed joints (MPa)

Welding Parameters		Tool offset (mm)		
		-1	0	+1
500 rpm	20 mm/min	108	130	65
600 rpm	20 mm/min	111	154	71

The increasing of tensile strength were due to better mixing of Al and Mg in welding zone and in top of that less welding defect in the stir zone. However, Firouzdour & Kou [19] have mentioned that there was not significantly different in tensile strength in constant rotation speed of 1400 rpm and welding travelling speed less than 75 mm/min when tool shifted into AZ31B and AZ31B was in advancing side.

A ratio between tensile strength of welded samples and ultimate stress of AZ31B base metal is called welding efficiency. The welding efficiency was varied from 29 % to 68 %. The minimum welding efficiency was related to sample in rotational speed of 500 rpm and pin at the -1 mm tool position which had lowest tensile strength and maximum efficiency was evaluated in sample when tool was in the middle in 600 rpm and 20 mm/min of welding traveling speed.

4. Conclusion

The following conclusions can be made based on the current study on FSW of dissimilar metals Mg-Al alloys:

1) Defect free AZ31B-Al6061(T6) dissimilar FSWed joints with good tensile strength was achieved with the combination of rotational speed of 500 and 600 rpm with travelling speed of 20 mm/min by locating Al on retreating side.

2) When tool positioned in -1 mm or +1 mm offsetting, some defects were found in SZ of Mg-Al dissimilar FSW joints such as cavity, tunnel, and crack. Also, weld defects were caused to reduce the mechanical properties of FSWed joints.

3) Mg side of SZ was separated into 3 regions; shoulder affected zone, banded zone and severe intercalated zone. The last two regions were in the middle and bottom of SZ which showed Mg-Al intercalated structures. Also, a thin and brittle layer of IMC was found in the interface of Mg-Al joints at stir zone.

4) Uniaxial tensile testing was carried out at room temperature and the ultimate tensile stress of welded joints was influenced by rotational speed and tool offsetting. Joint efficiency of the FSWed samples performed varied between 29% and 68% for the different parameters.

References

[1] Chen Y C and Nakata K 2008 Friction stir lap joining aluminum and magnesium alloys Scr. Mater. 58 433–6.

[2] McLean A A, Powell G L F, Brown I H and Linton V M 2003 Friction stir welding of magnesium alloy AZ31B to aluminium alloy 5083 Sci. Technol. Weld. Join. 8 462–4.

[3] Somasekharan A C and Murr L E 2004 Microstructures in friction-stir welded dissimilar magnesium alloys and magnesium alloys to 6061-T6 aluminum alloy Mater. Charact. 52 49–64.
[4] Fu B, Qin G, Li F, Meng X, Zhang J and Wu C 2015 Friction stir welding process of dissimilar metals of 6061-T6 aluminum alloy to AZ31B magnesium alloy J. Mater. Process. Technol. 218 38–47.
[5] Venkateswaran P, Xu Z-H, Li X and Reynolds A P 2009 Determination of mechanical properties of Al–Mg alloys dissimilar friction stir welded interface by indentation methods J. Mater. Sci. 44 4140–7.
[6] Dorbane A, Mansoor B, Ayoub G, Shunmugasamy V C and Imad A 2016 Mechanical, microstructural and fracture properties of dissimilar welds produced by friction stir welding of AZ31B and Al6061 Mater. Sci. Eng. A 651 720–33.
[7] Wang H-Y, Liu L-M, Zhu M-L and Wang H 2007 Laser weld bonding of A6061Al alloy to AZ31B Mg alloy Sci. Technol. Weld. Join. 12 261–5.
[8] Long T and Reynolds A P 2006 Parametric studies of friction stir welding by commercial fluid dynamics simulation Sci. Technol. Weld. Join. 11 200–8.
[9] Zhao L M and Zhang Z D 2008 Effect of Zn alloy interlayer on interface microstructure and strength of diffusion-bonded Mg-Al joints Scr. Mater. 58 283–6.
[10] Borrisutthekul R, Miyashita Y and Mutoh Y 2005 Dissimilar material laser welding between magnesium alloy AZ31B and aluminum alloy A5052-O Sci. Technol. Adv. Mater. 6 199–204.
[11] Venkateswaran P and Reynolds A P 2012 Factors affecting the properties of Friction Stir Welds between aluminum and magnesium alloys Mater. Sci. Eng. A 545 26–37.
[12] Somasekharan A C and Murr L E 2004 Microstructures in friction-stir welded dissimilar magnesium alloys and magnesium alloys to 6061-T6 aluminum alloy Mater. Charact. 52 49–64.
[13] Sato Y S, Park S H C, Michiuchi M and Kokawa H 2004 Constitutional liquation during dissimilar friction stir welding of Al and Mg alloys Scr. Mater. 50 1233–6.
[14] Chang W-S, Rajesh S R, Chun C-K and Kim H-J 2011 Microstructure and Mechanical Properties of Hybrid Laser-Friction Stir Welding between AA6061-T6 Al Alloy and AZ31 Mg Alloy J. Mater. Sci. Technol. 27 199–204.
[15] Venkateswaran P, Xu Z-H, Li X and Reynolds A P 2009 Determination of mechanical properties of Al–Mg alloys dissimilar friction stir welded interface by indentation methods J. Mater. Sci. 44 4140–7.
[16] Fu B, Qin G, Li F, Meng X, Zhang J and Wu C 2015 Friction stir welding process of dissimilar metals of 6061-T6 aluminum alloy to AZ31B magnesium alloy J. Mater. Process. Technol. 218 38–47.
[17] Kostka A, Coelho R S, dos Santos J and Pyzalla A R 2009 Microstructure of friction stir welding of aluminium alloy to magnesium alloy Scr. Mater. 60 953–6.
[18] Zhao Y, Lu Z, Yan K and Huang L 2015 Microstructural characterizations and mechanical properties in underwater friction stir welding of aluminum and magnesium dissimilar alloys Mater. Des. 65 675–81.
[19] Firouzdor V and Kou S 2010 Al-to-Mg Friction Stir Welding: Effect of Material Position, Travel Speed, and Rotation Speed Metall. Mater. Trans. A 41 2914–35.
[20] Firouzdor V and Kou S 2010 Formation of Liquid and Intermetallics in Al-to-Mg Friction Stir Welding Metall. Mater. Trans. A 41 3238–51.
[21] Kou S and Firouzdor V 2009 Al-to-Mg Friction Stir Welding : Effect of Positions of Al and Mg with Respect to the Welding Tool Weld. J. 213–24.
[22] Dorbane A, Mansoor B, Ayoub G, Shunmugasamy V C and Imad A 2016 Mechanical, microstructural and fracture properties of dissimilar welds produced by friction stir welding of AZ31B and Al6061 Mater. Sci. Eng. A 651 720–33.

[23] Masoudian A, Tahaei A, Shakiba A, Sharifianjazi F and Mohandesi J A 2014 Microstructure and mechanical properties of friction stir weld of dissimilar AZ31-O magnesium alloy to 6061-T6 aluminum alloy Trans. Nonferrous Met. Soc. China 24 1317–22.
[24] Morishige T, Kawaguchi A, Tsujikawa M, Hino M, Hirata T and Higashi K 2008 Dissimilar Welding of Al and Mg Alloys by FSW Mater. Trans. 49 1129–31.

IOP Conf. Series: Materials Science and Engineering **238** (2017) 012019 doi:10.1088/1757-899X/238/1/012019

Investigation of preheating method on joint strength of aluminium-stainless steel dissimilar welding using metal inert gas (MIG) process

M R Mohamad, L H Shah and M Ishak

Faculty of Mechanical Engineering, Universiti Malaysia Pahang, 26600 Pekan, Pahang, Malaysia

E-mail: luqmanhakim@ump.edu.my

Abstract. This study investigates AA6061-O and SUS304 dissimilar welding with preheating of stainless steel SUS304 prior to welding process. The welding method used was metal inert gas (MIG) with butt joint type weld. The mechanical strength was investigated using tensile test. Meanwhile, the macrostructure and microstructure of the specimens were analyzed using optical microscope, scanning electron microscopy (SEM) and energy dispersive spectroscopy (EDS). The tensile tests indicate that the preheated specimen with 90 °C have the maximum ultimate tensile strength of 111 MPa. In addition to that, the intermetallic compound (IMC) of the all the specimen was observed to be in the range of 1.59 µm to 10.8 µm. Fracture failures occur at the IMC interfaces on all specimen, where a thicker IMC layer consequently yields a lower tensile value. It can be concluded that the optimum parameters for AA6061-O to SUS304 welding can be achieved at 90 °C with 17.5 V welding voltage and 110 A welding current.

1. Introduction

Metal inert gas (MIG) welding is a prominent fusion welding method and is widely used in various industrial applications such as in shipbuilding, construction and automotive sector. Due to its long history, many researches have been done on its feasibility in various environments, and this is evident through the many international standards that have been established on MIG welding.

To meet new industrial demands, a new welding technique known as dissimilar welding was introduced. Such technique opens the possibilities to join metals of dissimilar alloy or material together to form a unique part. Dissimilar welding research have since seen many successful joining of different materials to aluminium, such as aluminium to steel [1], aluminium to copper [2], aluminium to magnesium [3] and aluminium to titanium [4].

Out of the many combinations, dissimilar welding between aluminium alloys and various steel alloys has been the most widely studied upon. This is primarily due to the high strength and toughness characteristics of steel alloys that compliments well with the low cost, light weight and corrosion resistant characteristic of aluminium alloys [5, 6]. However, such combination is difficult due to the near zero solubility and the stark contrast of the melting temperature between aluminium and steel. Consequently, this facilitates the formation of a notorious Fe_xAl_y brittle intermetallic (IMC) compound between the two metals during the joining process [7, 8]. Since these IMC layer formation is inevitable but adversely affects the weld strength, Kreimeyer and Sepold [9] have suggested that a weld joint with an IMC layer thickness of less than 10 µm is acceptable and can be considered mechanically sound.

To mitigate the formation of such brittle structure, several researches have proposed the use of preheating prior to the welding process. Recent researches by Kreimeyer and Vollertsen [10] and Bang et al. [11] have suggested a secondary heat source such as a tungsten inert gas (TIG) torch in order to preheat the steel counterpart prior to welding to enhance the joint strength. Other works also suggested the use of laser as a preheating source to enhance fusion and solid state welding between aluminium and steel [12–14]. These researches result in significant improvements to the joint strength due to the suppression of the deleterious IMC layer [15].

However, such additional preheating source set up is costly and complex. Shah et al. [16] conducted the preheating method using a mobile burner torch on tungsten inert gas (TIG) dissimilar welding of AA1100-SUS304 and AA6061-SUS304, respectively. The method proved to be simple, effective and cheap. Both results have shown an increase in the tensile strength of the joint when steel is preheated up to 150 °C.

2. Experimental Procedure

2.1 Materials

The base materials used in this study was a 2 mm thick aluminium alloy AA6061-O and stainless steel SUS304. Both metals were cut into plates with dimensions of 150×55×2 mm. The filler metals used for this study is ER4043. Table 1 shows the chemical composition (in wt%) for base metals AA6061 and SUS304 as well as filler material ER4043, respectively.

Table 1. Chemical composition (wt%) of AA6061, SUS304 and ER4043.

Materials	Zn	Ti	C	Mg	Si	Cr	Mn	Cu	Fe	Al
AA6061	0.021	0.028	-	0.775	0.643	0.191	0.073	0.236	0.303	Bal.
SUS304	-	-	0.06	-	0.397	17.90	1.010	0.210	Bal.	0.004
ER4043	0.1	0.2	-	0.05	<6.0	-	0.05	0.3	0.8	Bal.

2.2 Welding process

The joining process was carried out by using an automatic Dr Well DM-500EF MIG machine. The oxide layers on specimen surface were removed using a metal brush to prevent inclusion of impurities in the weld joint. They were then carefully cleaned using acetone and left to dry to remove dirt and grease.

Table 2. Fixed and varied MIG welding parameter

Welding parameter	Value
Welding speed	4 mm/sec
Preheating temperature	90 °C, 180 °C, 270 °C
Welding voltage	17 V, 17.5 V, 18 V
Welding current	105 A, 110 A, 115 A

Details of the fixed and varied welding parameters are presented in table 2. Through preliminary tests, no sound welds can be formed using lower or higher values of welding voltage and current. Welding with the parameters mentioned in table 2 without preheating also yielded unsuccessful joints. After the welding set up and just prior to the welding process, the stainless-steel counterpart was preheated using a commercially available gas burner torch up to the desired temperature. The temperature was monitored closely using a mobile Center 300 Series digital thermometer with surface probe. Once the desired temperature was reached, the welding process commenced.

2.3 Mechanical and microstructural analysis

In order to investigate the specimen joint strength, tensile tests were conducted using the Shimadzu Universal AG-X tensile test machine. The dimension for the tensile specimen was cut using an electrical discharge machining (EDM) wire cut according to the ASTM-E8 standard. Figure 1 shows the dimension (in millimeters) for the tensile specimen. The weld joint is to be located at the middle of the dog bone shape.

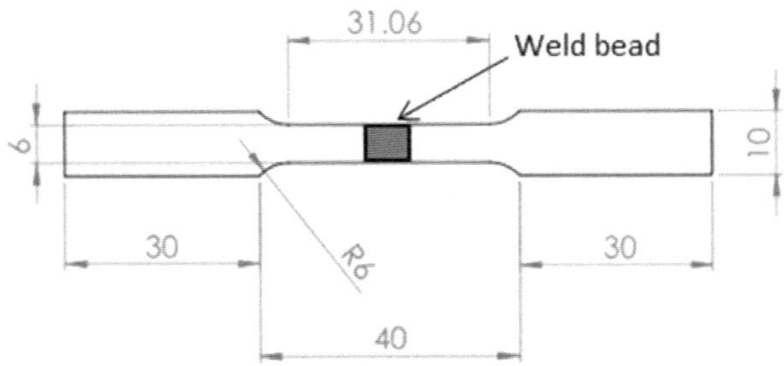

Figure 1. Tensile specimen dimensions (in mm).

For the microstructure analysis, the specimens were ground using 240, 320, 400 and 600 grit papers in that order and polished using 6 μm and 0.05 μm polishing plates to remove scratches on the surface to be observed. To reveal the grain boundaries, they are finally etched by dabbing a cotton bud immersed in an HF aqueous solution for the stainless-steel region, and a separate cotton bud immersed in Keller's reagent for the aluminium counterpart, respectively, in ambient temperature. Microstructure and elemental analysis of the specimen were observed using optical microscope, scanning electron microscopy (SEM) and energy dispersive X-ray spectroscopy (EDX).

3. Results and Discussion

3.1 Welding appearance

Figure 2 shows the best surface appearance of the welded specimens for each preheating temperature group. From the figure below, it can be seen that sound weld joints with minor defects can be obtained from the process. A better weld bead consistency can be observed from figure 2(a) using the 90 °C preheating temperature (sample 2), but better wetting can be observed for sample 4 and sample 9.

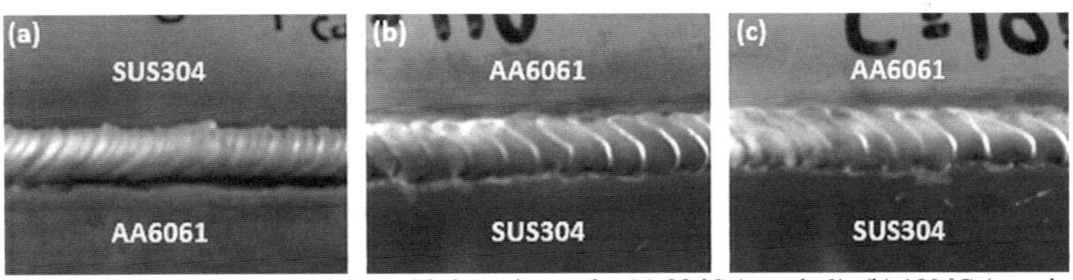

Figure 2. Surface appearance of the welded specimens for (a) 90 °C (sample 2), (b) 180 °C (sample 4) and (c) 270 °C (sample 9).

3.2 Tensile test

Figure 3 shows the side and top view of the fractured tensile test specimen. A more detail view of the weld bead before and after the fracture can be observed in figure 4. It can be seen in figure 4(a) that the welding process yielded a welding-brazing joint; the aluminium side combined with the aluminium filler to create a welding joint, whereas the ER4043 filler only brazed at the contact surface of the steel side. After fracture, figure 4(b) clearly shows that the line of fracture originates from the stainless steel and aluminium brazing boundary, i.e. the brittle intermetallic layer, indicating a brittle fracture mode.

The result of the tensile tests with the respective parameters is shown in table 3. The heat input from the MIG torch was calculated using the equation below [17]:

$$J = c \times UI \tag{1}$$

where J is the heat input (kJ/mm), c is the constant value of coefficient of welding efficiency over travel speed (mm/s), μ/V, U is the current (A) and I is the voltage (V).

IOP Conf. Series: Materials Science and Engineering **238** (2017) 012019 doi:10.1088/1757-899X/238/1/012019

From the data collected in the table, a graph of tensile strength versus preheating level and specimen number was plotted, as shown in figure 5. From the figure, it can be seen that the highest tensile strength is 111 MPa for sample 2 at preheating level 90 °C with 17.5 V welding voltage and 110 A welding current. This value is 90% of the tensile strength of the AA6061-O base metal (124 MPa). On the other hand, the lowest tensile strength value obtained was from sample 7 with 61 MPa using preheating temperature of 270 °C, voltage of 17 V and current of 110 A.

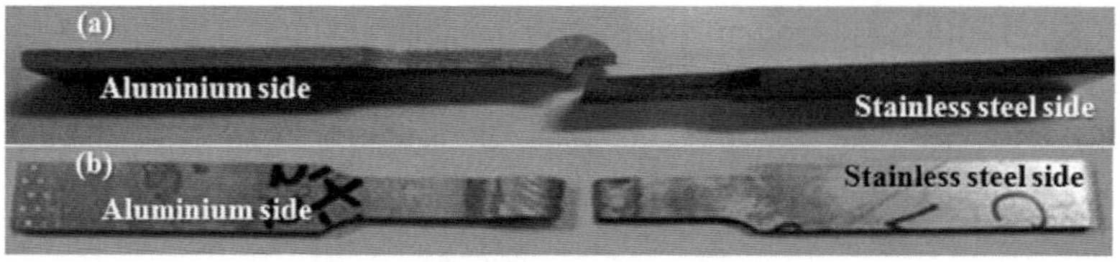

Figure 3. Fractured tensile specimen from the (a) side view, and (b) top view.

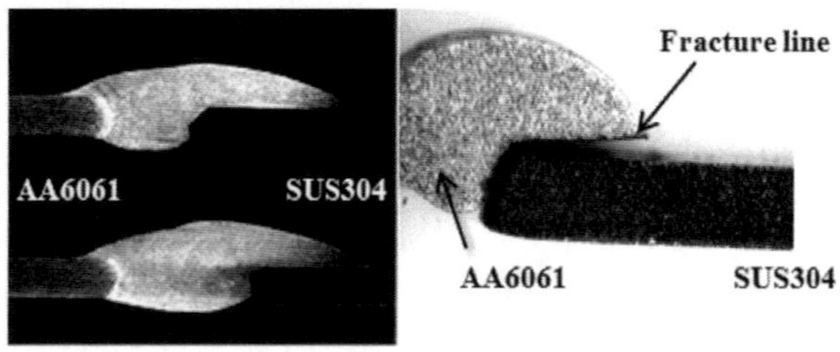

Figure 4. Cross sectional view of weld bead (a) as welded, and (b) after tensile testing.

Table 3. Tensile test results for all specimens.

Sample	Preheating(°C)	Voltage(V)	Current(C)	Heat input(kJ/mm)	Tensile strength(MPa)
1	90	17	105	1785*c*	104.53
2	90	17.5	110	1925*c*	111.27
3	90	18	115	2070*c*	66.46
4	180	17	115	1955*c*	95.04
5	180	17.5	105	1837.5*c*	88.34
6	180	18	110	1980*c*	89.28
7	270	17	110	1870*c*	60.53
8	270	17.5	115	2012.5*c*	80.97
9	270	18	105	1890*c*	82.67

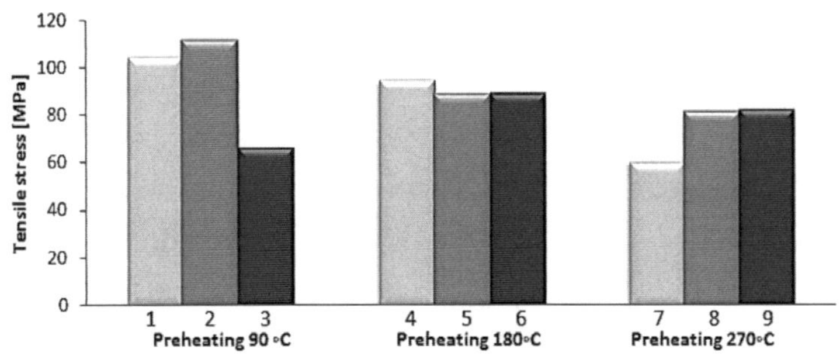

Figure 5. Tensile strength result.

3.3 Microstructure analysis

Figure 6 shows the microstructure morphology of the welded joint using an optical microscope at a low magnification (figure 6(a)-(c)) and at a high magnification (figure 6(d)-(f)), respectively. A mixed region can be observed between the weld zone and the aluminium base metal (figure 6(a)). On the other hand, a clear thin boundary between the weld zone and stainless steel base metal can be seen (figure 6(b)-(c)).

With a higher magnification, the transition between weld zone and heat affected zone (HAZ) of aluminium can be seen at the aluminium side (figure 6(d)). Coarse grains can be seen on the left side of figure, indicating that HAZ experienced thermal cycle without deformation. On the contrary, figure 6(e)-(f) show the existence of a thin inter metallic compound (IMC) layer on the inner and top faying surface of the steel sheet (inner and top IMC layer), respectively.

To better understand the IMC layer and the region surrounding it, the region is further examined using a scanning electron microscopy (SEM) and energy dispersive X-ray spectroscopy (EDX). Figure 7 shows the result of the SEM analysis at the inner side and top side of the bonded aluminium-steel surface. Due to the inconsistency of the thickness, several points were taken to determine the thickness range. The IMC thickness range for sample 2 was measured to be between 1.59 and 10.8 μm in size, indicating that a mechanically sound joint was achieved.

A line elemental analysis of the IMC layer region was conducted using EDX and its count-per-second versus position graphs of the main elements are shown in figure 8. From the individual elemental spectrums, it is evident that the left side of the analysis region (fig.8(a)) is dominated by aluminium and magnesium (fig.8(b)-(c)), whereas iron is mostly present on the right side (fig.8(d)). As for the IMC layer, it consists of aluminium, magnesium, silicon and carbon. Apart from the main elements of base metals, magnesium and silicon (fig.8(c) and (e)) are present primarily from the aluminium base metal and filler metal. However, since there is low carbon content in the initial set up, it is still unclear as to how carbon emerged in the IMC content. It can be assumed that the abrupt increase of carbon at the IMC layer (fig.8(f)) is primarily due to hydro carbon contaminants present at the faying surface of each material and at the filler metal surface. Further study is needed to clarify this issue.

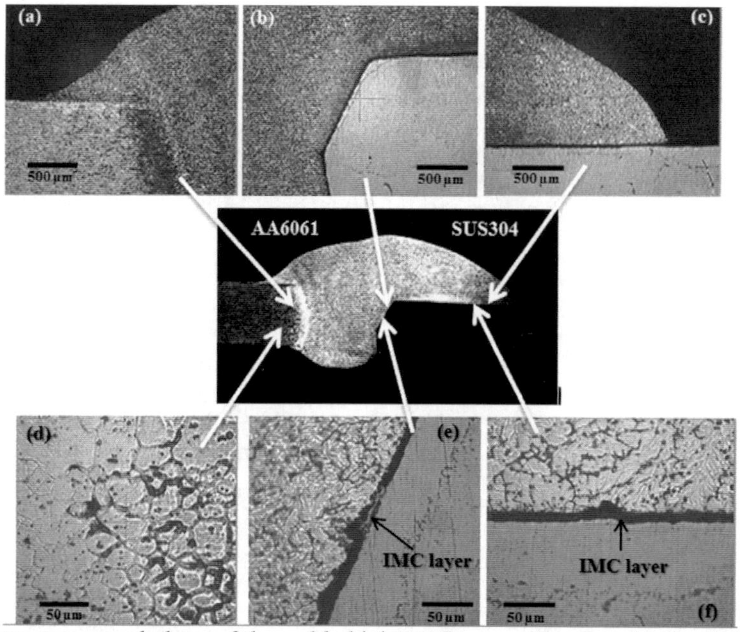

Figure 6. Microstructure morphology of the welded joint at 5× magnification; (a) aluminium joint, (b) inner steel joint, (c) top steel joint and at 50× magnification; (d) transition region of aluminium, (e) inner IMC layer, (f) top IMC layer.

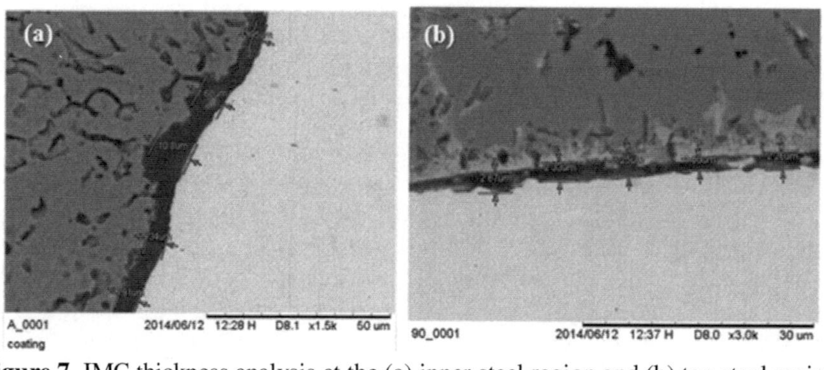

Figure 7. IMC thickness analysis at the (a) inner steel region and (b) top steel region.

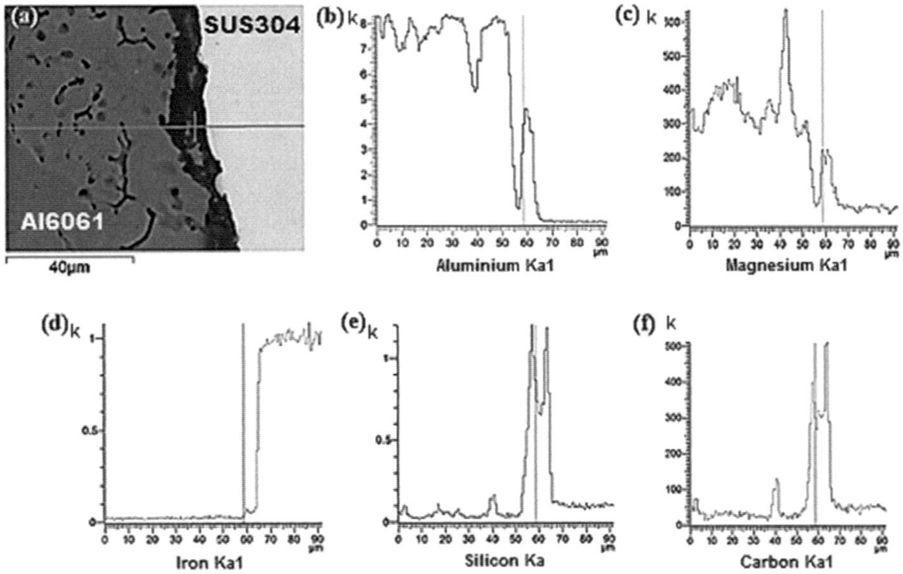

Figure 8. EDX line elemental analysis of the IMC region with individual elemental spectrums; (a)analyzed region; (b)aluminium; (c)iron; (d)magnesium; (e)silicon; (f)carbon.

The premise of the preheating method is to enhance diffusion between the base metals at a shorter duration and lower heat input. The IMC thickness under diffusion control process is governed by:

$$X = K\sqrt{t} \tag{2}$$

and

$$K = K_o \exp\left(-\frac{Q}{RT}\right) \tag{3}$$

where X is the IMC thickness (mm), t is the time (s) for diffusion, K_o is a constant, Q is the activation energy (J) for the growth of the layer, R is gas constant and T is the absolute temperature (K) [18, 19]. It is postulated that preheating on the stainless-steel counterpart provides additional energy to the system to surpass the threshold of Q in order to initiate IMC formation with lower heat input. With lower excessive heat input introduced to the system during the welding process, a shorter duration can be achieved, thus, from the equation (2), a thinner IMC can be formed.

The correlation between tensile strength, IMC thickness and the welding parameters can be explained as follows. As shown in figure 2, the samples with the best weld appearance have shown moderate heat input in the mid-range between the lowest and highest heat input, 1785c kJ/mm and 2070c kJ/mm, respectively. However, contrary to sample 2, tensile strength for both sample 4 and sample 9 show low tensile strength compared to other samples. This implies that despite the moderate heat input from the MIG torch which facilitates the wetting and consolidating process, the higher preheating temperature (180 °C and 270 °C respectively) of both samples may have caused the brittle IMC layers to grow thicker and cause detrimental effects on the samples' mechanical property. A general trend of decreasing tensile strength as preheating temperature increases can also be seen in figure 5, except for sample 3. Having the highest heat input (2070c kJ/mm), the excess heat input in sample 3 may have also caused further IMC growth and therefore a low tensile strength is produced.

4. Conclusion

Dissimilar metals joints between aluminum alloy AA6061-O and stainless steel SUS304 were successfully joined by MIG welding with preheating on the stainless-steel specimen before welding process. Major conclusions of this study could be summarized as follows:

i. All specimen yielded quality joints with minor defects.

ii. Fracture failure occurs at the IMC interface on all specimens. The preheated specimen with 90 °C preheating obtained the maximum ultimate tensile strength of 111 MPa.
iii. The IMC of the specimen with highest tensile strength was observed to be in the range of 1.59 μm to 10.8 μm.
iv. EDX analysis show that the IMC mainly consists of the base metal elements as well as contaminants.
v. The optimum parameters for dissimilar welding of AA6061-O and SUS304 can be achieved at 90 °C with 17.5 V welding voltage and 110 A welding current.

Acknowledgement
The authors would like to thank Universiti Malaysia Pahang for the research funding through RDU121409 and to the university staffs that have help to make this research a success.

References

[1] Rathod M J, Kutsuna M 2004 Joining of Aluminum Alloy 5052 and Low-Carbon Steel by Laser Roll Welding *J Weld* 83:16–26.
[2] Saeid T, Abdollah-zadeh A, Sazgari B 2010 Weldability and mechanical properties of dissimilar aluminum–copper lap joints made by friction stir welding *J Alloys Compd* 490:652–655.
[3] Kwon Y J, Shigematsu I, Saito N 2008 Dissimilar friction stir welding between magnesium and aluminum alloys *Mater Lett* 62:3827–3829.
[4] Dressler U, Biallas G, Alfaro Mercado U 2009 Friction stir welding of titanium alloy TiAl6V4 to aluminium alloy AA2024-T3 *Mater Sci Eng A* 526:113–117.
[5] Fukumoto S, Tsubakino H, Okita K 2000 Amorphization by friction welding between 5052 aluminum alloy and 304 stainless steel *Scr Mater* 42:807–812.
[6] Staubach M, Juttner S, Fussel U, Dietrich M 2008 Joining of steel-aluminium mixed joints with energy-reduced GMA processes and filler materials on an aluminium and zinc basis *Weld Cut* 7:30–38.
[7] Dharmendra C, Rao K P, Wilden J, Reich S 2011 Study on laser welding–brazing of zinc coated steel to aluminum alloy with a zinc based filler *Mater Sci Eng A* 528:1497–1503.
[8] Qiu R, Iwamoto C, Satonaka S 2009 The influence of reaction layer on the strength of aluminum/steel joint welded by resistance spot welding *Mater Charact* 60:156–159.
[9] Kreimeyer M, Sepold G 2002 Laser steel joined aluminium-hybrid structures. Proc. Int. Congr. Appl *Lasers Electro-Optics* 2002
[10] Kreimeyer M, Vollertsen F 2006 Gap tolerant joining of aluminium with steel sheets using the hybrid technique. In: Proc. Int. Congr. Appl. *Lasers Electro-Optics 2006*. Arizona, USA, pp 947–952
[11] Bang H, Bang H, Jeon G 2012 Gas tungsten arc welding assisted hybrid friction stir welding of dissimilar materials Al6061-T6 aluminum alloy and STS304 stainless steel. Mater Des 37:48–55.
[12] Qin G, Lei Z, Su Y 2014 Large spot laser assisted GMA brazing–fusion welding of aluminum alloy to galvanized steel *J Mater Process Technol* 214:2684–2692.
[13] Qin G, Su Y, Wang S 2014 Microstructures and properties of welded joint of aluminum alloy to galvanized steel by Nd:YAG laser + MIG arc hybrid brazing-fusion welding *Trans Nonferrous Met Soc China* 24:989–995.
[14] Merklein M, Giera A 2008 Laser assisted friction stir welding of drawable steel-aluminium tailored hybrids *Int J Mater Form* 1:1299–1302.
[15] Shah L H, Ishak M 2014 Review of Research Progress on Aluminum–Steel Dissimilar Welding *Mater Manuf Process* 29:928–933.
[16] Shah L H, Yusof A S M, Ishak M 2013 Investigation of heat treatment on weld joint quality of stainless steel and aluminium dissimilar welding *Proc. 16th Int. Conf. Adv. Mater. Process. Technol.*
[17] Nowacki J, Rybicki P 2005 The influence of welding heat input on submerged arc welded

duplex steel joints imperfections *J Mater Process Technol* 164–165:1082–1088.
[18] Kobayashi S, Yakou T 2002 Control of intermetallic compound layers at interface between steel and aluminum by diffusion-treatment *Mater Sci Eng A* 338:44–53.
[19] Borrisutthekul R, Yachi T, Miyashita Y, Mutoh Y 2007 Suppression of intermetallic reaction layer formation by controlling heat flow in dissimilar joining of steel and aluminum alloy. *Mater Sci Eng A* 467:108–113.

IOP Conf. Series: Materials Science and Engineering **238** (2017) 012020 doi:10.1088/1757-899X/238/1/012020

Effect of Copper-based Fillers Composition On Spreading and Wetting Behaviour

W N W M N Hissyam, A M Halil, T Kurniawan, M Ishak and T Ariga

Faculty of Mechanical Engineering, Universiti Malaysia Pahang, 26600, Pekan, Pahang, Malaysia

E-mail: aimanmh@ump.edu.my

Abstract. Wetting and spreading of molten brazing filler material are important factors that influence the brazing ability of a joint to be brazed. Several investigations into the wetting ability of a brazing filler alloy and its spreading area in the molten state, in addition to effects of brazing temperature on the contact angle, have been carried out. Generally, the composition of copper-based filler and temperature affect spreading of molten brazing filler material during brazing. Wetting by and interfacial reactions of the molten brazing filler material with the metallic substrate, especially, affect strongly the spreading of the filler material. In this study, the effects of filler composition and brazing temperature on the spreading of molten brazing filler metallic alloys were investigated. MBF 2005 (Cu, 5.7wt.%Ni, 9.7wt.%Sn, 7.0wt.%P), MBF 2002 (Cu, 9.9wt.%Ni, 4.0wt.%Sn, 7.8wt.%P) and VZ 2250 (Cu, 7.0wt.%Ni, 9.3wt.%Sn, 6.3wt.%P) alloys were used as brazing filler materials. Pure copper block and a rectangular plate were employed as the base metal. Brazing filler material and metallic base plate were first washed with acetone. Brazing was performed at 750°C and 800°C under Ar gas for 30 minutes using an electrically heated furnace, after which, the original spreading area, defined as the sessile drop area, and the apparent spreading area were both evaluated. It was observed that the spreading area and wetting angle influenced by the composition of copper-based filler.

1. Introduction

Brazing of copper joints is used extensively in the heat exchanger, EGR cooler, and dissimilar-metal joining. Studies found that the important factors for getting a good joint ability are influenced by the filler material properties, wetting and spreading by molten filler metal and the interfacial reactions between the molten filler and interfacial reactions between base materials and filler materials [1, 2]. A sound brazed joint generally results when an appropriate filler alloy is selected, surface cleanliness, metal and flux inclusions, different melting temperatures of the components, cooling rate of the joint, excessive alloying of the brazing metal, improper joint design, and tolerances [3]. Different percentage of the composition may influence the wetting ability of the filler on the base material. Fluxing is used except when an atmosphere is specifically introduced into the furnace to perform the same shielding gas. Furnaces are either batch or continuous types with possibly atmosphere controls and should have automatic time and temperature [6, 7].

Other than that, different values of temperature are investigated [4]. Temperature has big influences in microstructure and strength in brazing [8,9]. The higher the temperature gives the higher strength of joining [10]. These two factors will distribute the spreading area and the wetting angle of the molten fillers. High brazing temperatures inhibit the joining to the base metal, and the fluidity of brazing filler metals diminishes with increasing Cu content [5]. Apart from that, the joint strength will be affected by these two factors. Thus, the spreading area and wetting angle have been carried out to investigate the joint strength of the copper based fillers.

However, lack of studies explained the relationship between the individual element in fillers of copper brazing and the temperature of brazing. Thus, in this project, a study of the relationship of the elements in fillers and the temperature of brazing on filler spreading and wettability will be conducted.

2. Experimental Procedure

The specimens used was a copper plate and copper block. The copper plate was cut into rectangular shape and it was used for spreading method. The copper block was placed together into T-shape and it was used for the wetting method. The fillers for spreading method were cut at same weight which was approximately, 0.0040 g - 0.0042 g. The weight percentage of composition of the copper based fillers metals are listed in Table 1. The specimens were washed by acetone before it was placed into furnace brazing. The specimens were placed together in a ceramic boat into furnace brazing with the presence of argon gas as shielding gas to avoid oxidation at the flowrate of 1.0 l/min. The temperature used is 750°C and 800°C as manipulated parameter and the holding time of furnace brazing is 30 minutes. After the brazing was done, the specimens undergo the cold mounting process by using epoxy powder. Since the pure copper is soft, cold mounting is preferable than using hot mounting. After the cold mounting was fully cured, the specimens were grinded from 320 grit until 4000 grit. The specimens were grinded until the surface of wetting and spreading appears. Lastly, the specimens were polished by using the polishing pad until 0.5 μm grit. After polishing was done, the optical microscope was used to observed and captured the spreading of molten filler and the wetting of the filler. The method to calculate the spreading area was by using perimeter gridline. The wetting angle was found by using the method of right angled triangle.

Table 1. Chemical composition (wt. %) of the copper based filler metal.

Filler metal	Cu (wt.%)	Ni (wt.%)	Sn (wt.%)	P (wt.%)
MBF 2005	77.6	5.7	9.7	7.0
MBF 2002	78.3	9.9	4.0	7.8
VZ 2250	77.4	7.0	9.3	6.3

3. Results and Discussion

3.1. Spreading Area

Spreading test results are shown in Figure 1 and Figure 2. According to these results, filler material composition influences the original spreading areas and the temperature influences the spreading ability of the copper-based filler material.

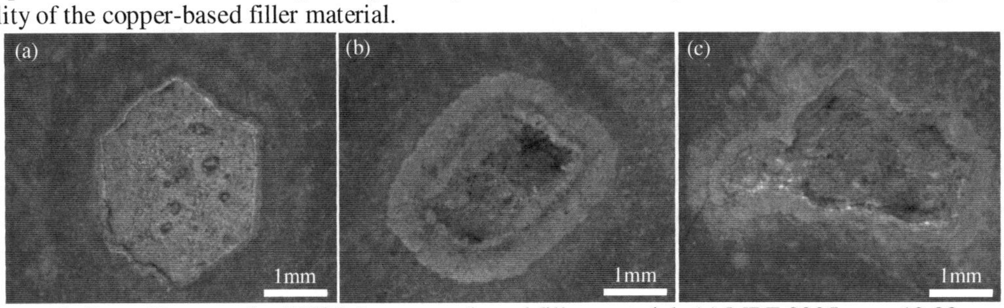

Figure 1. Spreading area for 750°C for copper-based filler material; (a) MBF 2005, area:12.88 mm², (b) MBF 2002, area:15.50 mm² and (c) VZ 2250, area:17.25 mm².

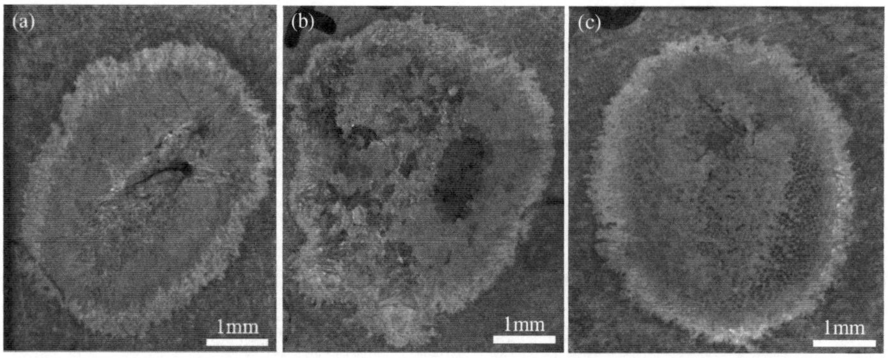

Figure 2. Spreading area for 800°C for copper-based filler material; (a) MBF 2005, area: 48.00 mm², (b) MBF 2002, area: 58.00 mm² and (c) VZ 2250, area: 61.00 mm².

From the figure 1, the result shows that MBF 2005 have the lowest spreading ability. The filler is fully deposited on the surface of base metal but still have a contact surface on the base metal. MBF 2002 also not completely molten but the spreading area of MBF 2002 is higher than MBF 2005. For VZ 2250, the molten part of the filler is lower than MBF 2002 but it has the largest spreading area. The deposited part of VZ 2250 is higher than MBF 2002 and MBF 2005. Compared with figure 1, figure 2 shows that all the fillers are molten at 800 °C. This is due to the composition of the filler which is VZ 2250 have quite balance percentage of Ni, Sn, and P. The balanced composition in VZ 2250 encourage the filler to provide highest spreading area. According to J. Li from his journal, contain higher Sn in filler composition will increase pitting for the joints but it inhibits the filler to spreading at the molten base [10]. It is shown in filler MBF 2005 at 800 °C which is, the higher the Sn composition will have lowest spreading area. MBF 2002 and VZ 2250 still have a small amount of deposited filler on the base metal. It is not completely molten due to the different percentage of filler composition. Thus, from both figure, higher or lower percentage of certain composition does not influence the ability to spread.

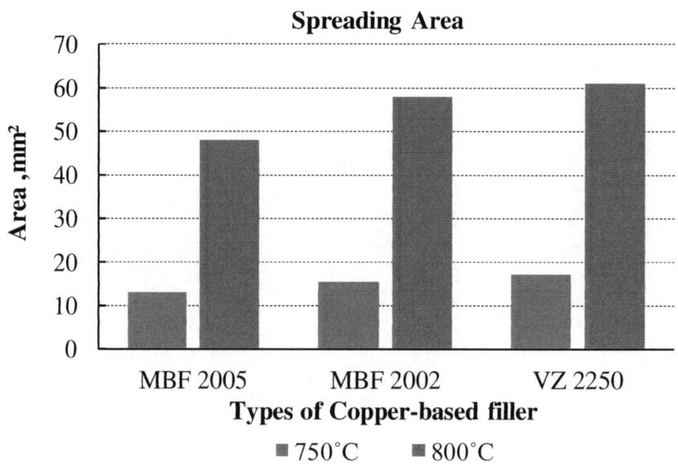

Figure 3. Spreading Area with Different Temperature.

Figure 3 shows the difference of spreading area between the temperature. According to the graph, we can relate that composition of the copper-based filler and the temperature of the furnace brazing give high influences to the spreading area of the molten filler. The higher the temperature of furnace brazing, the bigger the spreading area of the molten fillers. For the filler of MBF 2005, the highest percentage of other filler, it gives the lowest of the ability to spread on the base metal. The present of the highest percentage of P in MBF 2002, it increases the spreading ability of MBF2002 more than MBF 2005. Since the percentage of phosphorus in any particular alloy has a marked effect on its temperature (the temperature at which it becomes fully molten) and hence it's flow characteristics. The closer an alloy's phosphorus content is to the eutectic composition, the better its flow properties. In addition, small changes in phosphorus content can also have a marked effect on an alloy's flow properties. VZ 2250 has present of the phosphorous, it contributes the fluidity at a fast rate. An increase in phosphorous content increases the fluidity of the liquid metal. More fluid alloy requires a looser-fitting space to allow capillary attraction to retain the liquid. Thus, the higher the temperature, the bigger the spreading area.

3.2. Wetting Angle

Figure 4 and figure 5 shows the wetting angles of the copper-based filler materials with different temperatures. According to the results, there are slight differences in wetting angle of the filler. In other words, the temperature also gives influences to the wetting ability of the molten filler

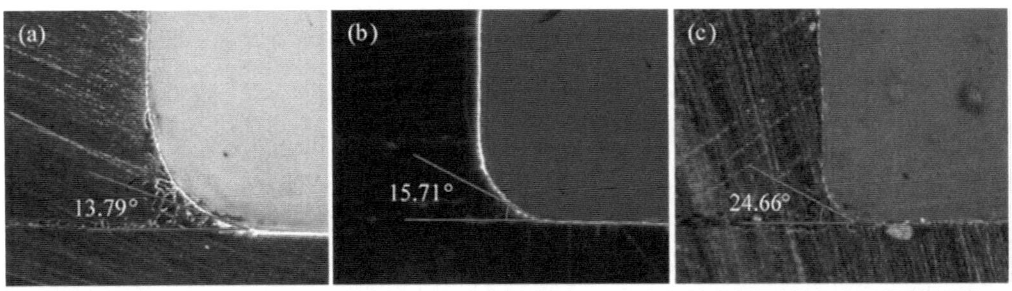

Figure 4. Wetting angle for 750°C for copper-based filler material; (a) MBF 2005 (b) MBF 2002 and (c) VZ 2250.

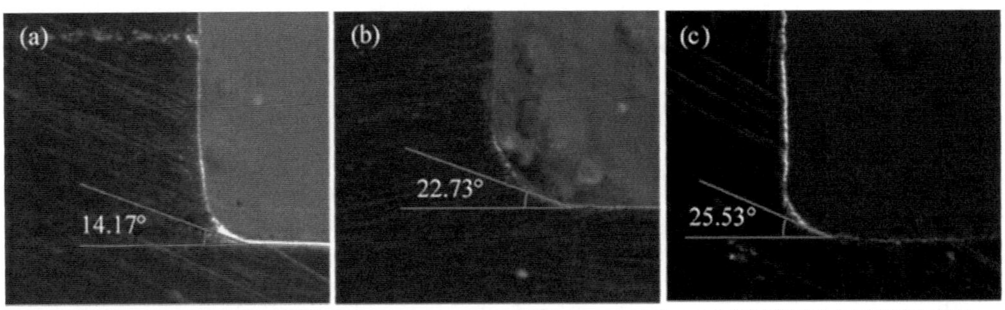

Figure 5. Wetting angle for 800°C for copper-based filler material; (a) MBF 2005 (b) MBF 2002 and (c) VZ 2250.

From the both figures above, the result shows that MBF 2005 have the lowest wetting angle and VZ 2250 has the higher wetting angle. Compared to the both temperature, the wetting angles slightly increase from 750°C to 800 °C. Due to different percentage composition of the filler, the resulting wetting angle also affected. MBF 2005 has higher contain of Sn but low contain of Ni among these three fillers but MBF 2002 has high contain of Nickel and very low contain of Sn. These shows that high or low in composition percentage do not give good wetting ability. Unlike VZ2250, it has quite a similar percentage of Ni, Sn and P composition. These differences in composition affects the wetting ability of the copper-based fillers.

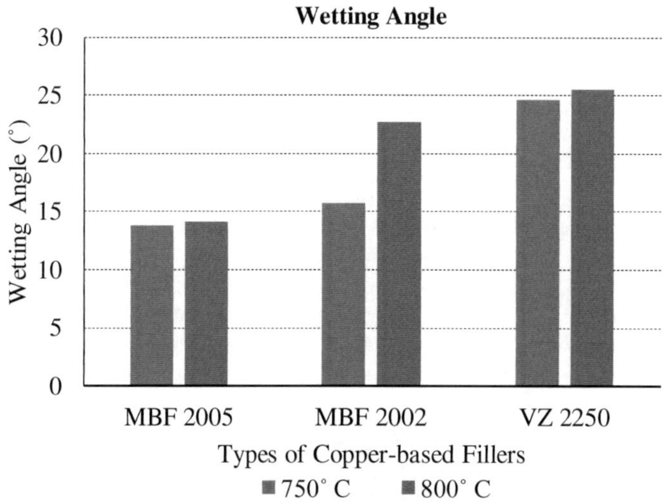

Figure 6. Wetting Angle with Different Temperature

From the figure above shows the relationship between wetting angle and temperature of furnace brazing. Wetting angle increase with increasing the temperature of the brazing. Since Nickle is to enhance the strength and hardness to the filler and Sn is to reduces friction and improves braze ability of the filler, thus the higher percentage of Ni and Sn will contribute the higher wetting angle of the copper-based fillers. The present of high Ni and Sn will inhibit the filler to get a higher wetting angle. VZ 2250 has quite balancing in composition Ni, Sn, and P contain. As shown in Table 1, the percentage of composition from copper-based fillers gives effect on wetting ability between fillers and base plate. This result shows that the balance percentage in the composition provides good wetting ability.

4. Conclusion

In this study, the effect of Copper-based fillers composition on spreading and wetting angle is investigated. The results as follows:

- The spreading area increased with increasing the brazing temperature
- The wetting angle is higher with increasing the brazing temperature
- The wetting and spreading are affected by the different percentage composition in copper-based filler metal. However, the pros and cons in the percentage of composition filler material will exhibit wetting and spreading ability. Thus, the same average of percentage in filler composition is needed to provide good wetting and spreading ability.

References

[1] American Welding Society C3 Committee on Brazing Soldering 2007 *Brazing Handbook Fifth Edition*, (AWS2007), Chapter1, 2

[2] Kozlova O, Voytovych R, Eustathopoulos N and Protsenko P J 2010 Wetting dynamics versus interfacial reactivity of AlSi alloys on carbon *Mater Sci* **45** 8 2150-2156

[3] Karamıs M B¸ Tasdemirci A and Nair F 2003 Microstructural Analysis and Discontinuities in the Brazed Zone of Copper Tubes *Journal of Materials Processing Technology* **141** 302-312

[4] Satoshi K, Tetsurou K, Yukiakira H and Yasuyuki M 2014 Wetting and Spreading Behavior of Molten Brazing Filler Metallic Alloys on Metallic Substrate *IOP Conf. Series: Materials Science and Engineering* **61** 012017

IOP Conf. Series: Materials Science and Engineering **238** (2017) 012020 doi:10.1088/1757-899X/238/1/012020

[5] Tatsuya F, Yuki W, Yasuyuki M and Fumio K 2014 Brazing of copper to stainless steel with a low-silver-content brazing filler metal *IOP Conf. Series: Materials Science and Engineering* **61** 012016

[6] Robert W M 2004 Joining of Materials and Structures *First published, the United States of America, Elsevier Butterworth–Heinemann* 349-387

[7] Flom Y and Wang L 2004 Flaw Tolerance in Lap Shear Brazed Joints - Part 1 *Welding Journal*, January 32-38

[8] Guedes A, Pinto A and Vieira M 1999 The effect of brazing temperature on the titanium/glass-ceramic bonding *Mater Process Technol* **102** 6 92-93

[9] Jinglong L, Zengchan H, Jiangtao X, Fusheng Z and Wen-Ya L 2009 Study on microstructure and strength of AgNi alloy/pure Al vacuum diffusion bonded joints *Mater Des* **30** 8 3265–8

[10] Li J, Liu Y, Tan Y, Li Y, Zhang L, Wu S and Jia P 2013 Effect of Tin Addition on Primary Silicon Recovery in Si-Al Melt During Solidification Refining of Silicon *J. Cryst. Growth* **371** 1-6